Lecture Notes in Economics and Mathematical Systems

Managing Editors: M. Beckmann and W. Krelle

300

Ulrich Derigs

Programming in Networks and Graphs

On the Combinatorial Background
and Near-Equivalence of
Network Flow and Matching Algorithms

Springer-Verlag Berlin Heidelberg GmbH

Author

Prof. Dr. Dr. Ulrich Derigs
Lehrstuhl für Betriebswirtschaftslehre, insbesondere Betriebsinformatik
Universität Bayreuth, Universitätsstr. 30, D-8580 Bayreuth, FRG

ISBN 978-3-540-18969-5   ISBN 978-3-642-51713-6 (eBook)
DOI 10.1007/978-3-642-51713-6

Library of Congress Cataloging-in-Publication Data. Derigs, Ulrich, 1950- Programming in networks and graphs. (Lecture notes in economics and mathematical systems; 300) Bibliography: p.
1. Combinatorial optimization. 2. Network analysis (Planning) 3. Matching theory. I. Title. II. Series.
QA402.5.D47 1988 511'.6 88-3163

2142/3140-543210

Meinen Eltern in Dankbarkeit gewidmet

C'est le temps que tu as perdu pour ta rose qui fait ta rose si importante.

(A. de Sainte Exupery)

Preface

Since the origins of combinatorial optimization network flow-, matching- and certain matroid problems have been treated as cornerstone problems of this discipline. For each of these problems a variety of algorithms was developed which were able to solve even large-scale, real-world problems. This last property made these classes outstanding and attractive.

During the seventies with the development of the "complexity theory" of computer algorithms the universe of combinatorial optimization problems was divided into two major classes: the "easy" problems (i.e. those for which a polynomial algorithm was known) and the "hard" problems (i.e. the so-called NP-complete problems which all become easy if the existence of a polynomial algorithm for only one representative can be shown).

The last break-through in this direction was done with the development of the ellipsoid-algorithm for linear programs and the understanding of its impact for combinatorial optimization by setting up a general ellipsoid-cutting-plane-scheme for combinatorial optimization problems. Yet although by this framework a number of special combinatorial problems could newly be classified to be easy and the borderline between "easy" and "hard" could be discovered almost completely, this result is not extremely satisfactory since the basic ingredients of this approach are by no means combinatorially natured - at least the combinatorial backbones have not been discovered so far. Moreover this scheme seems not at all to be practical in general.

Thus when looking for nice <u>combinatorial</u> algorithms which work in polynomial time one is basically back to the roots, but now knowing for certainty that

- network flow problems
- matching problems and

- matroid (intersection) problems

are the only major combinatorial optimization problems of practical relevance which are "easy" under this restriction.

Now there is an immediate interest in analysing and understanding the special charme or nature of these three problems which makes them "easy". The construction of a "super-problem" which contains all three problems as special cases and which itself is also "easy" would certainly help to decode this property.

Yet so far one did only succeed in constructing easy super-problems for any two of these three cornerstone-problems — thereby creating another three cornerstone-problems:

- the matroid matching problem

- the submodular flow problem

- the general matching problem in bidirected graphs.

In its general form, the matroid-intersection problem is of purely theoretical interest only and from a practical point of view network flow problems, and since only recently, matching problems, too, play the dominant role. Both problems were generalized to the so-called general matching problem or bidirected flow problem. Yet what has even more consequences is the fact that one could show that network flow problems can be reduced to a genuine matching problem. Thus we are really left with two cornerstone problems only.

In the literature "network flow" and "matching" are treated separately in general and for each class a variety of "different" algorithms has been developed up to now. In fact from an algorithmic point of view the development of matching procedures was influenced significantly by network flow techniques.

Knowing the relation among both problems one might expect a relation among these procedures, too. The question that we want to contribute to in this tract is concerned with the existence of a common combinatorial principle which might be inherent in all those at first sight different approaches.

In fact we will show that all common network flow and matching algorithms follow (implicitely) the "shortest augmenting path" concept which can be interpreted as a greedy-like decision rule where the optimal solution is built up through a sequence of "local optimal" solutions. Thereby polynomiality can only be guaranteed if this myopic decision rule is combined with an anticipant organisation.

Table of contents

PART I

PRELIMINARIES

Chapter 1. Terminology

In the following sections we introduce some concepts and definitions from linear algebra and graph theory which are frequently used all over the tract.

1.1. General Notation

By $\mathbb{R}(\mathbb{Q}, \mathbb{Z}, \mathbb{N})$ we denote the real (rational, integer, natural) numbers. The natural numbers $\mathbb{N}$ do no contain zero, hence we define $\mathbb{N}_0 := \mathbb{N} \cup \{0\}$. By $\mathbb{R}_+$ we denote the set of nonnegative reals. If E is a finite set then we use $\mathbb{R}^E$ to denote the set of $|E|$–tuples, with every component of a vector $x \in \mathbb{R}^E$ indexed by an element of E. Given a vector $x \in \mathbb{R}^E$ and $F \subseteq E$ we denote by $x \mid_F$ the restriction of x to the components indexed by elements from F. Given a vector $x \in \mathbb{R}^E$ and a subset $F \subseteq E$ we abbreviate

$$x(F) := \sum_{e \in F} x_e .$$

The inner product of two vectors $x, y \in \mathbb{R}^n$ is denoted by $x'y$. Any $|E|$–tuple $x \in \mathbb{R}^E$ can also be interpreted as a mapping $x : E \to \mathbb{R}$, then given another $|E|$–tuple $c = (c_e \mid e \in E)$ we sometimes write $c(x)$ instead of $c'x$.

A matrix $A \in \mathbb{R}^{m \times n}$ is said to be a (m, n)–matrix. Given two finite sets V and E, $\mathbb{R}^{V \times E}$ is the set of $(|V|, |E|)$–matrices with row index set V and column index set E. The inverse matrix of a matrix A - if it exists - is denoted by A^{-1}. Let I, J be finite sets. If $A \in \mathbb{R}^{I \times J}$ is the matrix $(a_{ij} \mid i \in I, j \in J)$, then for any $K \subseteq I$ we let A_K denote the matrix $(a_{ij} \mid i \in K, j \in J)$. If K is a single element j we abbreviate $A_{\{j\}}$ by A_j. With $a \in \mathbb{R}^n \backslash \{0\}, a_0 \in \mathbb{R}$ the set $\{x \in \mathbb{R}^n \mid a'x \leq a^0\}$ is called a halfspace of $\mathbb{R}^n$. A polyhedron is the intersection of finitely many halfspaces. A bounded polyhedron is called a polytope.

Throughout this study we use the convention that the minimum over the empty set is set to infinity. If x is a real number then $\lfloor x \rfloor$ and $\lceil x \rceil$ denote the lower integer part and the upper integer part of x , respectively.

1.2. Graphs

A graph $G = (V(G), E(G))$ consists of a finite nonempty set $V(G)$ of *nodes*
and a set $E(G)$ of edges which are unordered pairs of different nodes. An edge
$e \in E(G)$ joining or linking two nodes $i, j \in V(G)$ is denoted by $\{i, j\}$ and
the nodes i, j are called the endnodes of e . The number of nodes of G
is called the order of G . Two nodes which are joined by an edge are called
adjacent or neighbours. For $v \in V(G)$ we denote by $N_G(v)$ the set of neighbours
and for $W \subseteq V(G)$ let $N_G(W) := \cup_{v \in W} N_G(v) \backslash W$. The set of edges having
a node $v \in V$ as one of their endnodes is denoted by $\delta_G(v)$ and the number
$d_G(v) := |\delta_G(v)|$ is called the degree of node $v \in V$. For any $W \subseteq V(G)$ we
let $\delta_G(W)$ denote the *coboundary* of W i.e. the set of edges with exactly one
endnote in W :

$$\delta_G(W) = \{\{i, j\} \in E \mid |\{i, j\} \cap W| = 1\} .$$

A graph $H = (V(H), E(H))$ is called subgraph of G if $V(H) \subseteq V(G)$ and
$E(H) \subseteq E(G)$ holds.
(Note that we will write $G = (V, E)$ and omit the subscript, i.e. we write $\delta(v)$
instead of $\delta_G(v)$, if this can be done without loss of clarity.)
For $W \subseteq V$ we define $\gamma(W) := \{\{i, j\} \in E \mid i, j \in W\}$ and $G[W] = (W, \gamma(W))$
the subgraph induced by W .
A graph is called complete if every two nodes are joined by an edge. The (up
to isomorphism unique) complete graph of order n is denoted by K_n .

A graph whose node set V can be partitioned into two disjoint sets V_1, V_2
with $V_1 \cup V_2 = V$ such that no two nodes in V_1 and no two nodes in V_2 are
adjacent is called bipartite. If $|V_1| = m$ and $|V_2| = n$ and every node in V_1 is
joined with every node in V_2 , then G is called complete bipartite and it is

3

denoted by $K_{m,n}$.

A path P in $G = (V, E)$ from i_0 to i_n is a sequence

$$i_0, e_1, i_1, e_2, \ldots, e_n, i_n \quad \text{for some} \quad n \in \mathbb{N}_0$$

such that

$$i_k \in V \quad \text{for} \quad 0 \le k \le n$$

$$e_k = \{i_{k-1}, i_k\} \in E \quad \text{for} \quad 1 \le k \le n \ .$$

A cycle in G is a "path" with $i_0 = i_n$. We call n the length of P and we say that P is odd or even if n is odd or even. A path is called simple if no node occurs more than once. Then we write $P = (V(P), E(P))$ with $V(P) := \{i_k \mid k = 0, \ldots, n\}$ and $E(P) = \{e_k \mid k = 1, \ldots, n\}$. Yet we will also write $v \in P$ resp. $e \in P$ when referring to a node $v \in V(P)$ resp. edge $e \in E(P)$ if this can be done without loss of clarity.

A graph $G = (V, E)$ is connected if for every $\{i_1, i_2\} \subseteq V$ there is a path P in G from i_1 to i_2 . A maximal connected subgraph of G is called a component of G . A cutnode v of $G = (V, E)$ is a node $v \in V$ such that with $W = V \backslash \{v\}$, the graph $G[W]$ has more components than G . G is called nonseparable if G is connected and has no cutnode. An isthmus of $G = (V, E)$ is an edge $e \in E$ such that with $F = E \backslash \{e\}$ the graph $H = (V, F)$ has more components than G .

A polygon is a connected subgraph $P = (V(P), E(P))$ with $d_P(v) = 2$ for all $v \in V(P)$. If $|E(P)|$ is odd (even) we call P an odd (even) polygon. König [1936] has shown that a graph $G = (V, E)$ is bipartite iff G contains no odd polygon.

A forest is a graph which contains no polygon; a tree is a connected forest. Let $G = (V, E)$ and $F \subseteq E$ such that $T = (V, F)$ is a tree, then T is called a spanning tree of G . The following are well known properties of trees:

Every tree $T = (V(T), E(T))$ with $|V(T)| \ge 1$ has at least two nodes of degree 1.

4

If $T = (V(T), E(T))$ is a tree, then $|E(T)| = |V(T)| - 1$. Every component H of G contains a spanning tree.

A matching M in a graph $G = (V, E)$ is a set of edges such that no two edges of M have a common endnode. A matching M is called perfect if every node is contained in one edge of M.

The node–edge incidence matrix A of a graph $G = (V, E)$ is the matrix $A \in \{0, 1\}^{V \times E}$ where

$$a_{ve} = \begin{cases} 1 & \text{if } v \text{ is an endnode of } e \\ 0 & \text{otherwise .} \end{cases}$$

1.3. Digraphs

A directed graph or digraph $G = (V, E)$ consists of a finite nonempty set V of nodes and a set E of ordered pairs (i, j) of different elements of V. Elements of E are called directed edges or arcs. If $e = (i, j) \in E$ then e is said to go from i to j and the nodes i and j are called endnodes of e with i the tail of e and j the head of e. Given a directed edge $e \in E$ we write $h(e)$ and $t(e)$ for the head resp. tail of e.
For $i \in V$ we denote

$$\delta^+(i) := \{e \in E \mid h(e) = i\}$$
$$\delta^-(i) := \{e \in E \mid t(e) = i\}$$

and the numbers $d^+(i) := |\delta^+(i)|$ resp. $d^-(i) := |\delta^-(i)|$ are called the outdegree resp. indegree of i, while $d(i) := d^+(i) + d^-(i)$ is called the degree of i.
For any set $W \subseteq V$ we set

$$\delta^+(W) = \{(i, j) \in E \mid i \in W, j \notin W\}$$
$$\delta^-(W) = \{(i, j) \in E \mid i \notin W, j \in W\}$$

A path in a digraph $G = (V, E)$ is a sequence

$$i_0, e_1, i_1, \ldots, e_n, i_n \quad \text{with} \quad n \in \mathbb{N}_0 \quad \text{and}$$

$$i_k \in V \quad \text{for} \quad k = 0, 1, \ldots, n$$

$$e_k = (i_{k-1}, i_k) \in E \quad \text{for} \quad k = 1, \ldots, n \ .$$

A path is called simple if no node occurs more than once. The node–edge incidence matrix of a digraph $G = (V, E)$ is the matrix $A \in \{-1, 0, 1\}^{V \times E}$ where

$$a_{ve} = \begin{cases} 1 & \text{if} \quad v = h(e) \\ -1 & \text{if} \quad v = t(e) \\ 0 & \text{otherwise} \end{cases} \ .$$

Note that the above definitions of graphs and digraphs do not allow so–called loops and multiple edges. Those graphs are sometimes called simple graphs. In this study we will mainly work in simple graphs. Yet there are some occasions where nonsimple graphs are required. Given a digraph $G = (V, E)$ and $i \in V$ we call the "ordered pair" (i, i) a loop in G and analogously we can introduce loops in an (undirected) graph as an (undirected) edge with two identical endnodes. A multigraph is a "graph" in which some edges have the same endnodes. Those edges are called multiple edges or parallel edges.

Chapter 2. Linear Programming

In the following we shortly present some basic results from linear programming which are fundamental for solving network flow and matching problems. The concept we draw on most frequently concerns duality relations. As another prerequisite we outline the simplex method and the primal–dual concept from a viewpoint that is most convenient for our study of network flow and matching algorithms.

2.1. Important Primal–Dual Relations

Let I, J be finite sets and let $K \subseteq I$. Let $A \in \mathbb{R}^{I \times J}$, $b \in \mathbb{R}^I$ and $c \in \mathbb{R}^J$. Then a (primal) linear programming problem (LP) is

$$\text{minimize } c'x \quad \text{for } x \in \mathbb{R}^J \quad \text{satisfying}$$

$$A_K x = b_K$$

$$A_{I \setminus K} x \geq b_{I \setminus K}$$

$$x \geq 0$$

The dual linear program is the linear program

$$\text{maximize } b'y \quad \text{for } y \in \mathbb{R}^I \quad \text{satisfying}$$

$$A'y \leq c$$

$$y_{I \setminus K} \geq 0$$

$$y_K \text{ unrestricted in sign}$$

It can be shown that any primal LP of the above form can be transformed to a LP in which $K = I$.

A vector $x \in \mathbb{R}^J$ satisfying the constraints of the primal LP is called a feasible solution to the primal problem; dual feasibility of a vector $y \in \mathbb{R}^I$ is defined analogously. A primal feasible solution x which minimizes $c'x$ for all

7

feasible primal solutions is called an optimal primal solution; an optimal dual solution is defined analogously.

Now the following theorem is fundamental in LP–theory (cf. Dantzig [1966]).

Theorem 2.1.

For any linear program exactly one of the following situations occur:

(i)   *There exists no feasible solution,*

(ii)   *the problem is unbounded i.e. for every $c_0 \in \mathbb{R}$ there is a feasible x such that $c'x < c_0$,*

(iii) *there is an optimal feasible solution.*

Let us denote by P the set of feasible solutions for the primal LP, then P is a polyhedron and the following theorem holds.

Theorem 2.2.

If the primal linear program has an optimal solution, then it has an optimal solution which is a vertex of P .

The following theorems give the relationship between the objective function values for primal and dual feasible solutions. Proofs can be found in Dantzig [1963] and numerous other standard LP–textbooks.

Theorem 2.3.   *("Weak LP–duality theorem")*
If x is a feasible primal solution and y is a feasible dual solution then

$$c'x \geq b'y .$$

Corollary 2.4.

If the dual program is unbounded, i.e. for any $b_0 \in \mathbb{R}$ there exist feasible dual solutions y such that $b'y \geq b_0$, then the primal problem is infeasible, i.e. $P = \emptyset$.

Theorem 2.5. *("Strong LP–duality theorem")*

If the primal problem is feasible and bounded i.e. $P \neq \emptyset$ and exists $c_0 \in \mathbb{R}$ such that $c'x \geq c_0$ for all $x \in P$, then there is an optimal primal solution x^0 and an optimal dual solution y^0 and $c'x^0 = b'y^0$.

The following theorem is <u>the</u> tool to prove optimality of a pair (x^0, y^0) of primal resp. dual feasible solutions and it will be used extensively in this study.

Theorem 2.6. *("Complementary slackness theorem")*

A primal feasible solution x^0 and a dual feasible solution y^0 are optimal iff

$$x_j^0 > 0 \text{ implies } (A')_j y^0 = c_j \quad \text{for } j \in J$$
$$y_i^0 > 0 \text{ implies } A_i x^0 = b_i \quad \text{for } i \in I \backslash K$$

2.2. The Simplex–Method

In the following we describe the simplex method from a viewpoint that is most convenient for treating our kind of special structured (network flow and matching) problems. Using the notation of the last section we assume furtheron that our (primal) linear program is given in the form with $K = I$, i.e.

$$min \ c'x$$
$$Ax = b$$
$$x \geq 0$$

W.l.o.g. we assume that A is of full row rank and $I = \{1, \ldots, m\}$ and $J = \{1, \ldots, n\}$.

For $B \subseteq \{1, \ldots, n\}$ we consider the system

$$A^B x_B = b$$

$$x_{J \setminus B} = 0$$

If A^B is nonsingular then B is called a (column-) *basis* of A and the above system has the unique solution x with

$$x_B = (A^B)^{-1} b$$

$$x_{J \setminus B} = 0$$

the associated *basic solution*.

Now if $x_B \geq 0$, then x is (primal) feasible and B is called a *feasible basis*, while $x = \binom{x_B}{0}$ is called a feasible basic solution. We know that if the primal linear program has an optimal solution, then it has an optimal solution which is a basic solution.

Now consider the associated dual linear program

$$max \ b' y$$

$$A' y \leq c \ .$$

Given a basis B of A, let y_B be the unique solution of the system

$$(A^B)' y_B = c_B \ .$$

Then it is easy to see that

$$c' x = (c^B)' x_B = b' y_B$$

thus if y_B is dual feasible, then x is an optimal solution.

Now we define the reduced cost $\bar{c}_j(B)$ of each "column" $j \in \{1, \ldots, n\}$ as follows

$$\bar{c}_j(B) = c_j - A'_j y_B \ .$$

Then we call a column j "dual feasible" or "dual infeasible" according to whether $\bar{c}_j(B) \geq 0$ or $\bar{c}_j(B) < 0$.

Now given a basis $B \subseteq \{1, \ldots, n\}$, a *pivot* from B to B' is an exchange of $j \in B$ with $j' \notin B$ such that

$$B' = (B \backslash \{j\}) \cup \{j'\}$$

is a basis. Here we say that j is pivoted out of the basis and j' is pivoted into the basis.

A *simplex pivot* is a pivot from a feasible basis B to B' such that

$$\bar{c}_{j'}(B) < 0$$
$$\bar{c}_j(B') > 0 \quad \text{and}$$

B' is a feasible basis.

The simplex method is started with any feasible basis B . If $\bar{c}_k(B) \geq 0$ for all $k \in J \setminus B$ then B is optimal. Otherwise simplex pivots are iteratively applied until either optimality is reached or P can be shown to be unbounded.

In general the simplex method is described to work on the matrix A , or a related simplex–tableau, and general pivot–selection rules and transformations of the matrix (tableau) are given. We will not state these calculations here since we will apply the simplex method to special structured (graph–)problems only, and we will in these cases show how the simplex method can be interpreted to work on the graph/network rather than the linear system (matrix A).

Since the number of feasible bases of a linear system is finite the simplex method is finite unless it encounters a (degenerate) subsequence of infinite length because of repeated bases. This phenomenon is called *cycling*.

In the absence of cycling the algorithm can admit a degenerate sequence of exponential length in m which also prevents the simplex method from beeing a "good" algorithm in the sense of Edmonds' (cf. section 3.2.). This phenomenon is called *stalling*.

Although cycling is virtually unknown in practice several classes of LP's for which cycling occurs are presented in literature. Gassner [1964] has published the first example of cycling for network flow problems. Cycling can easily be avoided by introducing simple rules for choosing the entering column j' and the leaving column j ("pivoting rules"), see for example Dantzig, Orden and Wolfe [1955], Bland [1977]. Thus with these rules the simplex method becomes a finite method.

Avis and Chvatal [1978] have shown that if (besides cycling) stalling could be prevented for the simplex method then it becomes a good algorithm.

2.3. The Primal–Dual Method

In this section we present another general algorithm for solving LP's. It was first described by Dantzig, Ford and Fulkerson [1956] as a generalization of the Hungarian method for solving assignment problems developed by Kuhn [1955]. In fact this general approach can be viewed as the key idea within many "combinatorial" approaches for solving special "graph–related" optimization problems.

We start from the standard primal LP

$$\text{minimize } z = c'x \quad \text{subject to} \qquad\qquad (P)$$
$$Ax = b$$
$$x \geq 0$$

with $A \in \mathbb{R}^{m \times n}, b \in \mathbb{R}^m, c \in \mathbb{R}^n$ and the equalities multiplied such that $b \geq 0$ holds. Let us also assume $c \geq 0$, for the moment. Then the dual program is

$$\text{max } w = b'y \quad \text{subject to} \qquad\qquad (D)$$
$$A'y \leq c$$
$$y \quad \text{unrestricted in sign.}$$

12

The complementary slackness conditions give necessary and sufficient conditions for a pair (x, y) to be optimal:

$$y_i(A_i x - b_i) = 0 \quad \text{for } i = 1, \ldots, m$$
$$(c_j - A'_j y) x_j = 0 \quad \text{for } j = 1, \ldots, n$$

Now suppose we have a feasible dual solution y at hand — if $c \geq 0$ then $y \equiv 0$ is dual feasible for instance. With respect to y we define

$$J := \{j \mid A'_j y = c_j\}$$

the set of "admissible" columns of A. If we can find now a feasible primal solution x with

$$x_j = 0 \quad \text{for } j \notin J$$

then x (and y) are optimal.

Thus this fact amounts to searching for a $x \in \mathbb{R}^n$ fulfilling

$$\sum_{j \in J} a_{ij} \cdot x_j = b_i \quad \text{for } i = 1, \ldots, m$$
$$x_j \geq 0 \quad \text{for } j \in J$$
$$x_j = 0 \quad \text{for } j \notin J$$

This problem can be formulated as a new LP, called the *restricted primal problem (RP)*:

$$min \ \xi = \sum_{i=1}^m s_i \quad \text{subject to} \qquad (RP)$$
$$\sum_{j \in J} a_{ij} x_j + s_i = b_i \quad \text{for } i = 1, \ldots, m$$
$$x_j \geq 0 \quad \text{for } j \in J$$
$$s_i \geq 0 \quad \text{for } i = 1, \ldots, m$$

If the optimal solution $\bar{x}$ for RP has value $\xi_{OPT} = 0$, then $\bar{x}$ is optimal for P, too. If $\xi_{OPT} > 0$ then we have to consider the dual problem to RP:

13

$$max \ w = b'y \ \text{subject to} \qquad\qquad (DRP)$$

$$(A')_j y \leq 0 \ \text{ for } j \in J$$

$$y_i \leq 1 \ \text{ for } i = 1, \ldots, m$$

$$y \ \text{unrestricted in sign.}$$

Now assume $(\bar{x}, \bar{y})$ a pair of optimal solutions for RP resp. DRP fulfilling the associated complementary slackness conditions.

Using the initial dual solution y and $\bar{y}$ we now construct an "improved dual" solution y^* as a special linear combination of y and $\bar{y}$, i.e.

$$y^* = y + \Theta \bar{y} .$$

The objective function value of y^* can be calculated as

$$b'y^* = b'y + \Theta b' \bar{y}$$

Since $b'\bar{y} = \xi_{OPT} > 0$ holds, any choice of $\Theta > 0$ leads to an improved dual solution y^* and we would like to choose $\Theta > 0$ as large as possible. Yet to remain feasible we have to require

$$(A')_j y^* = (A')_j y + \Theta (A')_j \bar{y} \leq c_j \quad \text{for all } j = 1, \ldots, n .$$

We already know that $(A')_j \bar{y} \leq 0$ for $j \in J$ since $\bar{y}$ is feasible for DRP. Now if $(A')_j \bar{y} \leq 0$ for all $j \notin J$, too, we can increase Θ indefinitely and thereby obtain indefinitely large objective function values. In this case D is unbounded and hence P is infeasible. Now let $(A')_j \bar{y} > 0$ for some $j \notin J$. In this case the feasibility criterion becomes

$$(A')_j y + \Theta (A')_j \bar{y} \leq c_j \quad \text{for all } j \notin J \text{ with } (A')_j \bar{y} > 0$$

and the largest Θ that maintains feasibility for $y^* = y + \Theta \bar{y}$ is

$$\Theta^* = min \ \{(c_j - (A')_j y) \ / \ (A')_j \bar{y} \mid j \notin J, (A')_j \bar{y} > 0\}.$$

14

With this definition of y^* the new (dual) objective function value is

$$w^* = b'y + \Theta^* b'\bar{y} .$$

Now we would use $y := y^*$ as new initial dual solution, redefine the set J and repeat the procedure until either an optimal solution x is obtained or we can show that P is infeasible.

Obviously the restricted primal problems could be solved by the simplex method for instance which would give us together with the optimal primal solution $\bar{x}$ the complementary optimal dual solution $\bar{y}$.

Yet in many combinatorial applications the restricted problems RP are again combinatorially natured. In general RP is a simpler combinatorial problem than P and direct, combinatorially motivated procedures for RP are available.

Thus given a combinatorial optimization problem which can be described as a linear program, the primal–dual algorithm can be interpreted as a scheme where the given combinatorial problem is reduced to a sequence of "simpler" combinatorial optimization problems.

At each iteration of the primal–dual algorithm we can "restart" from the optimal solution obtained on the previous iteration. This is due to the fact that every column of the optimal basis of a RP remains admissible with respect to the improved dual solution.

Let the minimum in the calculation of Θ^* occur for $j = j_0$, then j_0 becomes admissible, i.e. a member of J , in the next iteration since

$$(A')_{j_0} y^* = (A')_{j_0} y + \Theta^* (A')_{j_0} \bar{y} = c_{j_0} .$$

Moreover the reduced cost of column j_0 with respect to the old dual solution is $- (A')_{j_0} \bar{y} < 0$, hence at least one pivot–operation is necessary when we start the new iteration from the old optimal basic solution.

If we think of RP as defined on all the columns of A plus some artificial columns associated with the slack-variables s_i, $i = 1,\ldots,m$, then we would move from one basis of this system to another. If RP is not degenerate, the value ξ_{OPT} would decrease monotonically with each iteration and thus no basis can be repeated. Hence the primal-dual method is finite.

In the case of degeneracy one has to guarantee finiteness by one of the common anti-cycling rules for the primal simplex method.

Chapter 3. Combinatorial Optimization

An *optimization problem* P is to find a best possible "structure" subject to several input–parameters which are described and whose values are left unspecified. If all these parameters are set to certain values we speak of an *instance* I of the problem P. In the above general not very illustrative definition we have carefully distinguished between a problem and an instance of a problem. Informally, a problem is a collection of similar instances.

More formally an instance of an optimization problem is a pair (S, c) where S is any set, the domain of *feasible solutions* and c is a mapping $c : S \rightarrow \mathbb{R}$, the so–called *objective function*. Then the problem is to find a set $S^* \in S$ with

$$c(S^*) \leq c(S) \text{ for all } S \in S .$$

Such a set S^* is called *optimal solution*.

(Note that a more general framework is obtained when allowing functions $c : S \rightarrow H$ with $(H, \leq)$ a totally ordered set.)

An optimization problem is called a *discrete optimization problem* if for its instances (S, c), the feasible region S is a finite set. Thus discrete optimization problems can be formulated as

$$min \ \{c(S) \mid S \in S\}$$

For discrete optimization problems the existence of an optimal solution is guaranteed if $S \neq \emptyset$. In order to provide the above problem with more algebraic structure we assume for *combinatorial optimization problems* that the set of feasible solutions S is a subset of the power set of a given finite set E and that c is a separable function $c : E \rightarrow \mathbb{R}$ with

$$c(S) := \sum_{e \in S} c(e) .$$

Here the set E may be the set of edges of a graph $G = (V, E)$ and the set S may contain all those subsets of E defining a perfect matching in

G . Then the problem is called the min–cost–perfect–matching problem and an instance of this problem consists of a concrete graph $G = (V, E)$ and a cost–vector $(c_e \mid e \in E)$. Edmonds [1971] has called problems of the above form LOCO–problems (*linear objective combinatorial optimization problems*) which is more precise since also other objective funktions like bottleneck–, time–cost– and multicriteria–functions are relevant in connection with combinatorial "structures". Yet since in this study we only consider the most important objective function — the linear objective function — we will omit the prefix "linear objective" furtheron.

3.1. Polyhedral Combinatorics

Given an instance of a combinatorial optimization problem we can introduce for every subset $F \subseteq E$ an *incidence vector* $x^F \in \{0,1\}^E$ by setting

$$x_e^F = \begin{cases} 1, & \text{if } e \in F \\ 0, & \text{if } e \notin F \end{cases}$$

Defining $X := \{x^S \mid S \in \mathcal{S}\}$ we can rewrite P as a *boolean* or $(0,1)$ *optimization problem*

$$min \; \{\sum_{e \in E} c_e x_e \mid x \in X\} \; .$$

For most of the standard problems, the set $X \subseteq \{0,1\}^E$ can be represented as the set of integer solutions of a system of linear inequalities i.e.

$$X = \{x \in \mathbb{R}^n \mid Ax \geq b, \; x \geq 0, \; x \text{ integer valued .}\}$$

For example, given a graph $G = (V, E)$ and $X \subseteq \{0,1\}^E$ the set of incidence vectors of perfect matchings in G , we know that

$$X = \{x \in \mathbb{R}^E \mid Ax = 1, \; x \text{ integer valued}\}$$

where A is the $(|V|, |E|)$–node–edge incidence matrix of G .

Problems of the type

$$min \ \{c'x \mid Ax \geq b, \ x \geq 0, \ x \ \text{integer valued}\}$$

are called *integer linear programs*.

With any instance of an integer linear program P resp. its set X of feasible solutions we can associate two polytopes

$$\mathcal{F}(X) := \{x \in \mathbb{R}^E \mid Ax \geq b, \ x \geq 0\} \quad \text{and}$$

$$\mathcal{L}(X) := \text{conv} \ (X) \quad \text{the } convex \ hull \text{ of } X.$$

These polytopes give rise to two linear programs
the LP–*relaxation* $\mathcal{F}(P)$ of P :

$$min \ \{c'x \mid x \in \mathcal{F}(X)\}$$

and the LP–*formulation* $\mathcal{L}(P)$ of P :

$$min \ \{c'x \mid x \in \mathcal{L}(X)\} \ .$$

Let x_L resp. x_F be optimal solutions for $\mathcal{L}(P)$ resp. $\mathcal{F}(P)$, then the following relations hold

(i) $c'x_F \leq c'x_L$

(ii) $x_F \in X \Rightarrow x_F$ optimal for $\mathcal{L}(P)$.

Since the optimum of a linear program is always obtained at a vertex of the underlying polytope the following relation holds:

(iii)   Any optimal (vertex–) solution of $\mathcal{L}(P)$ is optimal for P , too.

The above representation of $\mathcal{L}(P)$ is not suitable for applying the effective LP–instruments like the simplex–method, the optimality conditions from duality theory etc. LP–techniques expect a description of the polytope in form of an

19

(in)equality system. Yet, due to a theorem of Weyl [1935], we know that for every polyhedron $\mathcal{L}$ represented as the convex hull of a set of points in $\mathbb{R}^E$ there exists a $(m, |E|)$–matrix D, $m \in \mathbb{N}$, and a vector $d \in \mathbb{R}^m$ such that

$$\mathcal{L} = \{x \in \mathbb{R}^E \mid Dx \geq d, \ x \geq 0\} \ .$$

Then the system (D, d) is called a linear description of $\mathcal{L}$ (resp. a linear description of problem P.) Unfortunately the proof techniques of the theorem of Weyl — although constructive — do not lead to "handy" linear descriptions of $\mathcal{L}$, usually.

Polyhedral combinatorics deals with the application of the theory of polyhedra and linear systems to combinatorics, and the determination of (complete and nonredundant) systems of equations and inequalities defining polyhedra associated with combinatorial optimization problems is one of the main topics of this discipline. An excellent introduction into the theory of polyhedra is Bachem and Grötschel [1982], while Pulleyblank [1983] gives an excellent survey on "polyhedral combinatorics" (see also Grötschel [1977]).

The following theorem is basic for the procedural establishment of linear systems defining polyhedra associated with combinatorial optimization problems.

Theorem 3.1.

Let $X \subseteq \mathbb{Z}^n$ finite, then $\mathcal{L} = \{x \in \mathbb{R}^n \mid Dx \geq d, \ x \geq 0\}$ *is the convex hull of* X *provided*

(i) $X \subseteq \mathcal{L}$ *and*

(ii) *for any* $c \in \mathbb{R}^n$ *the problem*

$$min \ \{c'x \mid Dx \geq d, \ x \geq 0\}$$

has an optimal solution $x^* \in X$.

Using LP–duality, Edmonds [1965] has given a fundamental procedure to prove that a system (D, d) is a linear description of $\mathcal{L}(X)$.

Recall that the dual program of

$$min \ \{c'x \mid Dx \geq d, \ x \geq 0\}$$

is given by

$$max \ \{d'y \mid D'y \leq c, \ y \geq 0\}$$

Now it suffices to give an algorithm which for every instance of P produces a pair (x^*, y^*) of feasible solution to the above dual pair of problems fulfilling complementary slackness. From that property follows the validity of the system (D, d).

Finally we are interested in the question whether exist combinatorial optimization problems P for which $\mathcal{F}(X) = \mathcal{L}(X)$ holds, i.e. which are essentially (simple) linear programming problems.

A square, integer matrix B is called *unimodular* if its determinant $\det(B) = \pm 1$ and an integer matrix A is called *totally unimodular* if every nonsingular square submatrix of A is unimodular.

The following theorem shows the importance of totally unimodular matrices in connection with integer linear programming.

Theorem 3.2. *(Hoffman and Kruskal [1956])*
If A is totally unimodular then all the vertices of the polyhedra

$$P^=(A, b) : = \{x \mid Ax = b, \ x \geq 0\}$$
$$P^\leq(A, b) : = \{x \mid Ax \leq b, \ x \geq 0\}$$

are integer for any integer vector b.

Thus an integer linear program with a constraint matrix A which is totally unimodular and an integer right hand side can be solved by LP–techniques, i.e.

the simplex–method, the primal–dual method etc. What is even more important is the fact that an optimal solution can be characterized by the existence of a complementary dual solution.

Unfortunately no "handy" characterization of totally unimodular matrices is known (cf. Camion [1965], Padberg [1976]). A rather useful sufficient (yet not necessary) condition is given in the next theorem which is due to Heller and Tompkins [1956].

Theorem 3.3

An integer matrix $A \in \{-1, 0, +1\}^{m \times n}$ is totally unimodular if every column contains at most two nonzero entries and the rows of A can be partitioned into two classes R_1 and R_2 such that

(i)   *if a column has two nonzero entries of the same sign, the associated rows are in different classes,*

(ii)   *if a column has two nonzero entries of different signs, the associated rows are in the same class.*

Due to the above theorem
— the node–edge incidence matrix of a digraph
— the node–edge incidence matrix of a bipartite graph
is totally unimodular.

Thus any LP whose constraint matrix A is one of those two kinds has only integer optimal basic solutions. Important cases are the shortest path problem, the min–cost–flow problem and the bipartite matching problem.

3.2. Algorithms and Complexity

In this work we study the combinatorial background of matching algorithms. Intuitively an algorithm is simply a sequence of instructions that solves any instances of a computational problem. Several precise mathematical definitions of the concept of algorithm have been proposed in literature. An excellent introduction into "algorithmic computation theory" with a thorough description of mathematical formalisms like "computer", "(decision)problem" and "algorithm" is Bachem [1982].

Like any combinatorial optimization problem the matching problem could for instance be solved by enumerating all possible solutions, comparing their "weight" and successively storing the best solution. This approach although obviously finite is not a practical method since the number of feasible solutions (matchings) grows rapidly with the size of the graph and thus we may not be able to enumerate all matchings within reasonable time.

The most common and most widely accepted performance measure for an algorithm is the time it spends before producing the final answer. Yet this amount of time may vary from one computer to another. In solving a problem an algorithm will perform a certain number of *elementary steps* — i.e. steps which do not depend on the size of the instance, for example arithmetic operations, comparisons of two numbers, checking whether two nodes are adjacent in a graph etc.

In the analysis of algorithms in this monograph we express the time requirements of an algorithm in terms of the number of elementary steps required for the execution of the algorithm on a hypothetical computer. Thereby we additionally make the "fixed–word" assumption, namely that the time required to perform arithmetic operations on two numbers is independent of the number of digits of these numbers.

Now it is evident that the running time of a specific algorithm to "solve" instances of a specific problem will depend on the "size" of this instance, resp. the size of the input for the algorithm and even for instances of the same size running time will vary due to different input data.

In the following we assume that the size of an instance can be expressed by one single number $n \in \mathbb{N}$ for all problems P . So for problems on graphs, the order of the graph is a useful measure.

Then the common simplification to measure the performance of an algorithm is to consider all instances of a given size and from this set to select the instance which has the worst performance rate. This approach measures the *worst–case–behaviour* of the algorithm and therefore yields a rather pessimistic (and often not practical) estimation about the expected "running time". Then the *time–complexity* of an algorithm is given as a function of the size of the input $f : \mathbb{N} \to \mathbb{N}$ by measuring the largest number of elementary steps necessary to solve an instance of given input size $n \in \mathbb{N}$.

Moreover we are interested in the *rate of growth* of the function f only and not in the exact function. To deal with rates of growth of functions the following definition is useful.

Let $f(n)$, $g(n)$ be two functions from the positive integers to the positive reals then we write $f(n) = O(g(n))$ if exists a constant $c > 0$ such that for large n, $f(n) \leq c \cdot g(n)$ and we say $f(n)$ *is of order* $g(n)$.
In measuring the complexity of algorithms we use "simple" functions $g(n)$ to bound the true complexity–function f from above. If for instance g is a polynomial of order k we will write $f(n) = O(n^k)$ neglecting lower order terms and all scalars.

We have introduced the complexity of an algorithm as a function of the input–size of the problem. Here the size of the input is defined to be the length of the associated sequence of symbols in an encoding of the data for a computer.

In this monograph we study problems on graphs. Thus the input is just the graph with the edge and node weights. A useful way to represent a graph is by its adjacency list where we record for each node v the set $N(v) \subseteq V$ of nodes adjacent to v . Now the complete list contains $2 \cdot |E|$ elements and hence the graph structure can be encoded in $O(|E|)$ space. The (integer) edge and node weights are represented in binary arithmetic and in this system the number of symbols to represent an integer n is $\lceil log_2 n \rceil$ (where $\lceil x \rceil$ denotes the smallest integer q with $q \geq x$).

Thus given a b–matching problem for instance with b_{max} the largest node weight, the space for the encoding of the node weights has length $O(|V| \cdot log_2 b_{max})$.

Now an algorithm is considered to be practically useful if its complexity grows *polynomially* with respect to the size of the input or is bounded by such a polynomial. Edmonds [1965] was the first who realised the significance of polynomially bounded algorithms in contrast to nonpolynomially algorithms. He called a polynomial algorithm "good" and a class of problems to be "well-solved" if a polynomial algorithm for solving problems out of this class is known.

To clearify the notion of polynomiality further we want to consider bMP and 1MP again. The 1–matching problem is well solved since several "different" good algorithms of complexity $O(|V|^3)$ are known. As we will see bMP can be reduced to a 1MP on a graph with $\sum_{v \in V} b_v$ nodes. Now applying the 1MP–algorithm to this instance gives an algorithm of complexity $O((\sum_{v \in V} b_v)^3) = O(((|V| \cdot b_{max})^3)$. But this is not a good algorithm since the input size of a bMP is $O(|E| + |V| \cdot log \ b_{max})$.

PART II

THE CLASS OF GENERAL MATCHING PROBLEMS

Network flow and matching problems are celebrated cornerstone problems in combinatorial optimization, integer programming and graph theory. In this part we show that both problems have a common "superproblem". Thus a unifying theory exists and common algorithmic principles can be expected. Then we illustrate the richness of this class by listing several important "special cases" which are standard problems in the OR–literature and which all have a specific charme. Thereafter we will establish a hierarchy among these special problems and we will crystalize several "key problems" within this class, which already "cover" the whole class from an algorithmic point of view, i.e. if these problems can be solved (by a polynomial algorithm) then all problems in this class can be solved (by a polynomial algorithm).

The most interesting result is thereby the fact that "network flow" can be interpreted as a special case of "matching" and in fact as the combinatorially less complex case of "bipartite matching".

Chapter 4. Three Cornerstone Problems

In this chapter we introduce the network flow problem and the graph matching problem as two special integer programs and we shortly discuss the relevance of these standard problems in Operations Research. A closer look at the structure of the constraint matrix of both problems leads to a common generalization: the general matching problem in bidirected graphs.

4.1. Network Flow

Let $G = (V, E)$ be a digraph representing a transportation network and let $b : V \to \mathbb{Z}$ be a demand/supply function on the set of nodes. By this mapping the nodes are classified into three different types:

- a node $v \in V$ with $b_v < 0$ is called an *origin* or *source node* i.e. a point where $(-b_v)$ units of good are available (stored),

- a node $v \in V$ with $b_v > 0$ is called a *destination* or *sink node* i.e. a point where b_v units of good are requested,

- a node $v \in V$ with $b_v = 0$ is called a *transhipment node*.

The task is now to fill the demand at the destinations by transporting the goods from the sources to the sinks along (directed) edges of G only.

If we denote by x_e the number of units transported on edge $e \in E$, then we have to find a mapping $x : E \to \mathbb{N}_o$ s.t.

$$x(\delta^+(v)) - x(\delta^-(v)) = b_v \qquad \text{for all } v \in V.$$

These conditions are called *flow conservation equalities* and a mapping fulfilling all these conditions is called a *flow* on G.

The problem is complicated by "capacity conditions" and "lower bounds" on the edges, i.e. for every edge $e \in E$ there are two numbers $l_e, u_e \in \mathbb{N}_o$ with

$l_e \leq u_e$, which give the minimal necessary usage and maximal possible usage of a single edge in the network. Then a mapping $x : E \to \mathbb{N}_o$ is called a *feasible flow* on G if it is a flow on G and

$$l_e \leq x_e \leq u_e \qquad \text{for all } e \in E.$$

Now the optimization problem occurs when a cost-function $c : E \to \mathbb{R}$ is introduced giving for every edge $e \in E$ the cost for transporting one unit from the tail node of e to the head node of e. Then we want to find a feasible flow x which minimizes total transportation cost $c(x)$ where

$$c(x) := \sum_{e \in E} c_e \cdot x_e$$

The above introduced *min-cost-network flow problem* (MCFP) can be formulated in matrix notation using the node-edge incidence matrix A of the graph G as an integer program as follows:

$$\begin{aligned} min \ \ &c'x & (MCFP) \\ &Ax = b \\ &l \leq x \leq u \\ &x \ \text{ integer valued.} \end{aligned}$$

Due to the "total unimodularity" property of A the integrality condition is superflous and MCFP is actually a linear program. Yet we will not focus on this property at this point. MCFP is one of the most celebrated problems in Mathematical Programming and Operations Research. Kennington and Helgason [1980] give the following list of systems in which network models have been used in practice:

- Production-distribution systems
- military logistic systems
- urban traffic systems
- railway systems
- communication systems
- pipe network systems
- routing and scheduling systems
- electrical networks.

Charnes et al. [1975] claim that about 70% of the real world mathematical programming problems consist of - or can be transformed into - network problems and they saw the following reasons for this large concentration on problems of the above kind, particularly in applications:

(i)   MCFP is accepted by business executives since the model is simple and illustrative,

(ii) answers to smaller problems can be calculated and controlled by hand,

(iii) computer codes are available since the early 50's, rather efficient ones since the early 70's at least.

Another important aspect is that network flow models can be used to approximate more general linear programs or even hard integer linear programs (cf. Mulvey [1978]).

To our opinion especially (iii) has impact on the concentration on network flow problems in practice and explains the large percentage estimated by Charnes et al. Practioner's - always under the pressure to present good (not necessarily optimal) solutions within short time - tend to apply models and technology which has already shown to be useful, has been tested already and is "available" (software). At least they use such successful applications as orientation. Since the "networkers" have been very active in reporting successful applications in more and more areas, practitioners were really lead into that direction of thinking during the process of modelling.

Efficient algorithms can only be developed if the problem has been analysed theoretically before. In the 50's (and 60's) network programming as a special linear program took benefit from any improvement in linear programming - on the other hand it also paid back and could influence the algorithmic development for general linear programming, too, as the example of the primal-dual approach demonstrates. Parallel to that process network problems were analysed from a more combinatorially point of view which immediately led to procedures shown to outperform LP–based approaches. The real break - through with respect to practical efficiency of network optimization codes occured in the early 70's when the LP-approach (network simplex method) was interpreted to operate combinatorially on the network rather than arithmetically on matrices/tableaus and highly efficient data-structures for handling those combinatorial or graphical structures were developed. Since that time a tremendous number of computational studies always presenting improved network codes has been published - by showing or sometimes even only "announcing" new advanced more clever strategies and data-structures. In the very last years the spectator of this race (practioner or researcher) could see the partisan fights among different research groups cool down. This indicates to some extent that the algorithmic development for network flow problems has reached a level where no significant improvements are possible or to be expected (unless somebody finds the next "big shot" and the race is restarted).

The state-of-the-art in network optimization is incorporated in quite a number of software-packages which are available (in principle) and which are able to solve also large-scale real world problems.

4.2. Graph Matching

Let $G = (V, E)$ be a graph representing a communication network, i.e. the nodes v can be viewed as terminals and there is an edge $\{v, w\}$ if two terminals can directly communicate. Every node v has a capacity/demand $b_v \in \mathbb{N}$ of possible information exchange operations. Now the task is to find an interconnection of the terminals such that the demand is fulfilled.

If we denote by x_e the number of uses of edge e, then we have to find a mapping $x : E \rightarrow \mathbb{N}_o$ such that

$$x(\delta(v)) = b_v \qquad \text{for all} \ \ v \in V .$$

These conditions are called *node constraints* and a mapping fulfilling all these conditions is called a *b-matching* on G. Now the problem may again be complicated by "capacity constraints" and "lower bounds" on the edges, i.e. for every edge $e \in E$ there are two numbers $l_e, u_e \in \mathbb{N}_o$ with $l_e \leq u_e$, which give the minimal necessary usage and maximal possible usage of a single edge in the graph. Then a mapping x is called a (feasible) *capacitated b-matching* if it is a b-matching on G and

$$l_e \leq x_e \leq u_e \qquad \text{for all} \ \ e \in E.$$

As in the network flow problem we assume a cost-function $c : E \rightarrow \mathbb{R}$ giving for every edge e the cost for exchanging one unit. Then the optimization problem is to find a capacitated b-matching which minimizes total communication cost $c(x)$ where

$$c(x) := \sum_{e \in E} c_e \cdot x_e.$$

The above introduced *min-cost-capacitated b-matching problem (CbMP)* can be formulated as an integer linear program in matrix notation using the node-edge incidence matrix A of the graph G

$$min\ c'x \qquad\qquad (CbMP)$$

$$Ax = b$$

$$l \leq x \leq u$$

$$x \quad \text{integer value}.$$

Although CbMP looks quite similar to MCFP the literature on this problem is by far not as numerous as for MCFP. Especially applications of this model are reported rather rarely. CbMP has found nearly no attention in the Operations Research literature. The first "Integer Programming Bibliography" compiled 1976 at the Institute für Ókonometrie und Operations Research, University of Bonn (cf. Kastning [1976]) contained already about 650 entries, i.e. articles, classified under the subject title "Network Flow", while only 50 articles were listed under the subject title "Matching". Since then the number of matching-related publications has increased steadily, but the proportion to "Network Flow" literature has not changed much.

As one can see from the above introduction "matchings" are as model as simple and illustrative as "network flows". Thus following Charnes' arguments the reason for "matchings" not being discovered to be a useful modeling tool may be found in the following facts.

(i) Even small problems cannot be solved or controlled by hand.

(ii) Nearly no computer software is available. Efficient codes have been produced only since very few years and only for special classes of matching problems.

Another reason could also be that

(iii) the matching model is unappropriate for real world problems.

Yet the very few real world matching-applications reported in the literature already demonstrate that the last reason is not a point at all. To our opinion the field of "matching-applications" will grow rapidly in the near future. "Matchings" are in a certain sense a rich and very attractive structure to model certain combinatorial problems (cf. Ball et al. [1983]). In Derigs [1982] a list of quite diverse "matching applications" can be found. Also matchings can be used in relaxation techniques for general 0–1–programs (cf. Weber [1978], Nemhauser and Weber [1979].)

To our opinion the lack of efficient matching software is the profound reason for the nonobservance of matchings by practioners. An illustrative example is the following: In 1976, Hasselström discovered that a standard problem within the process of urban transportation planning - the "optimization of route connections" - can be modeled as a matching problem. Yet since at that time no efficient matching code was available the problem was "solved" by a partial enumeration technique. (cf. Hasselström [1976], [1981]). Recently, with the aid of the author, an efficient matching code was developed for this special problem and implemented within this complex planning system resp. software package for urban transportation planning. This change did reduce the amount of computer time for that special module from several hours to about one minute. Moreover the quality of the solutions was increased significantly since now an *optimal* solution could be produced. (Note that during the planning process the matching problem has to be solved hundreds of times for graphs with about 40-80 nodes.)

The lack of practicable matching software lies primarily in the fact that matchings are a combinatorially more complex structure than "networks flows" - we will prove this statement lateron. Although the b-matching problem appeared in the literature already in the late 50's and early 60's (cf. Berge [1962]), the first efficient - non enumerative - algorithm for solving the least complex genuine matching problem was developed in the mid 60's (cf. Edmonds [1965]). Yet it took more than another ten years until the appropriate data-structures were developed, which led to applicable 1-matching codes. Even now the development of clever data-structures and updating formulas for matching algorithms is a vital topic in Computer Science and Mathematical Programming.

In contrast to "network flow", which we feel is already an established and extensively discovered field, the area of "matchings" has still a lot of "white spots" and we still can expect essential new ideas which will have impact on the practicability of matchings. We feel that progress with respect to computational aspects of matchings can only be achieved if the combinatorial background or the "backbone" of algorithmic principles is adequately understood. It is this question to which this work wants to contribute.

4.3. Programming in Bidirected Graphs

The min cost flow problem and the min-cost-capacitated b-matching problem constitute two special classes in Integer Programming which can be solved efficiently by exploiting their special combinatorial structure. (We want to mention only that both problems can also be solved "efficiently" by the polynomial ellipsoid-method of linear programming (cf. Khachian [1979], Grötschel et. al [1981]), yet this approach does not exploit the combinatorial nature of the problems and seems at least today and in its "pure" form not to be a serious alternative to the combinatorial approaches.)

In matrix notation MCFP and CbMP are of the same form where in both cases the constraint matrix A is the node-edge incidence matrix associated with the underlying graph resp. digraph. A closer look at the structure of those matrices A shows the following common properties:

- the elements of A are from the set $\{-1, 0, 1\}$,

- there are (exactly) two nonzero elements in every column of A.

Now the following two questions arise immediately.

(i) Does there exist a class of matrices containing the classes of node-edge incidence matrices of graphs and digraphs for which the associated integer program can be solved efficiently?

(ii) Does there exist a common generalization of MCFP and CbMP to a combinatorial or graphical problem, which allows good (combinatorial natured) algorithms?

These questions were treated and completely answered by Edmonds and Johnson [1970] who could show that the following class of matrices defines efficiently solvable integer linear programs.

A (m,n) matrix $A = (a_{ij})$ is contained in this class if

- the elements a_{ij} are drawn from the set $\{-2,-1,0,1,2\}$,

- every column contains at most two nonzero entries,

- $\sum_{i=1...m} |a_{ij}| \leq 2$ for all $j = 1,...,n$.

We will call these properties the "Edmonds-Johnson properties". It is evident that node-edge incidence matrices of graphs and digraphs are contained within this class. Now matrices of the above kind can be interpreted in a graphical environment, which leads to the definition of *bidirected graphs*.

A *bidirected graph* $G = (V, E)$ is a "graph" in which some edges (called *loops*) may have both ends incident with the same node, some edges (called *lobes*) may have only one end, and in which each edge has a direction associated with each node with which it is incident. Thus each edge has either one or two (not necessarily distinct) ends, and each of these ends is either a *tail end* or a *head end*.

With respect to a given set $W \subseteq V$ of nodes resp. a single node $v \in V$ we define for bidirected graphs

36

$\delta(W) :=$ the set of edges having exactly one end in W

$\gamma(W) :=$ the set of edges having both ends in W

$\delta^+(W) :=$ the set of edges in $\delta(W)$ whose single end in W is a head end

$\delta^-(W) :=$ the set of edges in $\delta(W)$ whose single end in W is a tail end

$\gamma^+(v) :=$ the set of loops having two head ends at v

$\gamma^-(v) :=$ the set of loops having two tail ends at v.

Now a bidirected graph $G = (V, E)$ can be represented in the following way by means of a node-edge-incidence matrix $A = (a_{ve}) \in \{-2, -1, 0, 1, 2\}^{V z E}$

where

$$
a_{ve} = \begin{cases}
-2 & \text{if } e \text{ is a loop with two tail ends at } v \\
-1 & \text{if } e \text{ is not a loop and has one tail end at } v \\
0 & \text{if if } e \text{ is not incident with } v \\
1 & \text{if } e \text{ is not a loop and has one head end at } v \\
2 & \text{if } e \text{ is a loop with two head ends at } v
\end{cases}
$$

(Remark: This representation does not allow for loops having one tail end and one head end. Yet in our applications of bidirected graphs it can be assumed that no such loops exist.)

It is easy to see that for the incidence matrix of a bidirected graph every column contains at most two nonzero entries and

$$
\sum_{v \in V} |a_{ve}| = \begin{cases} 1 & \text{if } e \text{ is a lobe} \\ 2 & \text{if } e \text{ is not a lobe.} \end{cases}
$$

On the other hand every matrix A having the Edmonds-Johnson properties leads to a proper definition of a bidirected graph (where every loop has two head ends or two tail ends).

The following example illustrates this relationship.

Example: Assume the $(3, 8)$ matrix

$$A = \begin{pmatrix} -1 & -2 & -1 & 1 & 0 & 0 & 0 & 0 \\ 0 & 0 & -1 & 0 & 2 & 1 & -1 & 1 \\ 0 & 0 & 0 & -1 & 0 & 0 & 0 & 1 \end{pmatrix}.$$

This matrix leads to the following bidirected graph $G = (V, E)$ with $V = \{v_1, v_2, v_3\}$ and $E = \{e_1, ..., e_8\}$

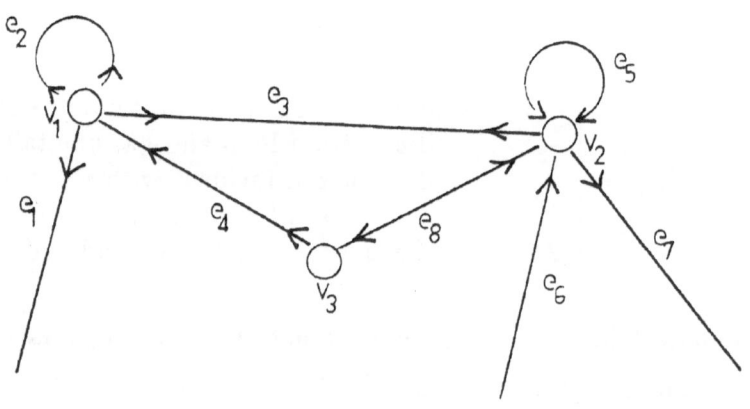

Figure 4.1. Bidirected graph

Now the integer program

$$min \; c'x \qquad s.t.$$

$$Ax = b$$

$$l \le x \le u$$

$$x \quad \text{integer valued}$$

with A a matrix fulfilling the Edmonds-Johnson properties can be formulated on the associated bidirected graph $G = (V, E)$ as follows:

$$min \sum_{e \in E} c_e \cdot x_e \quad subject \quad to$$

$$x(\delta^+(v)) - x(\delta^-(v)) + 2x(\gamma^+(v)) - 2x(\gamma^-(v)) = b_v \quad for \quad v \in V$$

$$l_e \le x_e \le u_e \quad for \quad e \in E$$

$$x \quad integer\ valued$$

In the literature the above problem is either referred to as the *bidirected flow problem* (Lawler [1976]) or the *general matching problem* (Araoz et al. [1983]) on G. Here the name depends on whether the author is approaching the problem from network flow theory or from matching theory. Because of reasons made clear in the sequel we will refer to this problem as a "matching problem". In fact one can show that every problem of the above type can be reduced to an equivalent "real" matching problem, i.e. a graph matching problem defined on a related graph. A similar reduction to a "pure" network problem is not known.

Thus a matching in a bidirected graph $G = (V, E)$ is an assignment of nonnegative integers x_e to the edges e of G such that x_e fulfills the "lower bound" and "capacity" conditions and the "demand" b_v at each node $v \in V$ is fulfilled, in the sense that edge e contributes $-2x_e, -x_e, 0, x_e, 2x_e$ units to the demand at v according to whether e has 2 tails, 1 tail, 0 ends, 1 head, 2 heads at v.

Now it is clear, too, why we can assume that G does not contain loops with one head end and one tail end. Such a loop e would not contribute to the demand at the node v with which it is incident. More precisely

- if $c_e < 0$ and the problem has a feasible solution then $x_e = u_e$ in any optimal solution,

- if $c_e \ge 0$ and the problem has a feasible solution then the problem has an optimal solution with $x_e = 0$.

In the following section we will demonstrate the richness of this class of combinatorial optimization problems by introducing a variety of classical and important special subclasses, i.e. well known standard optimization problems within this class of general matching problems.

Then we will show some reductions among several subclasses, thereby constituting a *hierarchy* among the problems of this class, and we will extract certain "key problems" which will be discussed in more detail furtheron.

Chapter 5. Important Subclasses

In the following we introduce a list of important combinatorial optimization problems which can be viewed as special cases (i.e. subproblems) of the general matching problem (resp. bidirected flow problem) and which have found special consideration in the literature - and which all show a special charme.

In fact we follow two lines of specialization

- the "network flow line" where we deal with directed graphs and

- the "matching line" where we deal with (undirected) graphs.

In the last section of this chapter we will present reductions among these classes and crystalize a set of "key problems".

5.1. Some Special Network Flow Problems

Here we assume $G = (V, E)$ a directed graph and

$c : E \to \mathbb{R}_+$	a cost function on the set of edges,
$l : E \to \mathbb{N}_0$	lower bounds on the set of edges,
$u : E \to \mathbb{N} \cup \{\infty\}$	upper bounds on the set of edges,
$b : E \to \mathbb{Z}$	a capacity function on the set of nodes

(N1) The min-cost flow problem (MCFP)

This problem has already been introduced in section (4.1.). Yet for reasons of completeness we give the formulation of this basic problem here again

$$min \sum_{e \in E} c_e x_e \quad \text{subject to}$$

$$x(\delta^+(v)) - x(\delta^-(v)) = b_v \quad \text{for } v \in V$$

$$l_e \leq x_e \leq u_e \quad \text{for } e \in E$$

$$x \quad \text{integer valued}$$

41

(N2) The circulation problem (CP)

This is the special case of a MCFP with

$$b_v = 0 \qquad \text{for all} \quad v \in V$$

i.e. all nodes are transhipment nodes.

(N3) The min–cost–(s, t)–flow–problem $(MC(s, t)FP)$

This is the problem

$$min \ \sum_{e \in E} c_e \cdot x_e$$

$$l_e \leq x_e \leq u_e \quad \text{for} \quad e \in E$$

$$x \ \text{integer valued}$$

with $s \in V$ the only source and $t \in V$ the only sink in the network. Here z is the maximal possible amount of flow from s to t in the network or any required level of flow to be sent from s to t .

So far we have introduced the circulation problem and the min–cost–(s, t)–flow problem as special cases of MCFP. Yet by some simple transformations it can be shown that all these problems are equivalent.

First we may assume w.l.o.g. that the underlying graph for MCFP has only one source, s say, and one sink, t say. Otherwise let

$$V^+ := \{v \in V \mid b_v > 0\} \subset V \qquad \text{the set of sinks and}$$
$$V^- := \{v \in V \mid b_v < 0\} \subset V \qquad \text{the set of sources.}$$

Then we introduce a dummy source $s \notin V$ and a dummy sink $t \notin V$ as well as artificial edges:

$$(s, v) \quad \text{with} \quad l_{sv} = 0 \ \text{and} \ u_{sv} = -b_v \quad \text{for all} \ v \in V^-$$
$$(v, t) \quad \text{with} \quad l_{tv} = 0 \ \text{and} \ u_{tv} = b_v \qquad \text{for all} \ v \in V^+.$$

and we define new node–capacities on $V' := V \cup \{s, t\}$

$$
b'_v := \begin{cases}
\displaystyle\sum_{v \in V^-} b_v & \text{for } v = s \\[2mm]
0 & \text{for } v \in V \\[2mm]
\displaystyle\sum_{v \in V^+} b_v & \text{for } v = t
\end{cases}
$$

Then the original MCFP on G is equivalent to the MCFP on $G' = (V', E')$, which is to find a least–cost s–t–flow of level b_t .

Now this problem can easily be transformed into an equivalent CP by adding a "return" edge (s, t) with $l_{ts} := u_{ts} := b_t$ and $c_{ts} := 0$.

If a min–cost–(s, t)–flow of maximum possible level is required we would define $l_{ts} := 0$, $u_{ts} := \infty$ and $c_{ts} := -M$ with M a sufficiently large number to ensure that x_{ts} becomes as large as possible.

Before we introduce more special subclasses of MCFP we will introduce a further reduction (which is nothing but a common trick in linear programming). Any MCFP with lower bounds $l_e > 0$ for $e \in E$ can be transformed into an equivalent MCFP with zero lower bounds:
Substituting x by $\tilde{x}$ where $\tilde{x}_e := x_e - l_e$ for $e \in E$, MCFP becomes

$$
min \ \sum_{e \in E} c_e \cdot \tilde{x}_e + \sum_{e \in E} c_e \cdot l_e \quad \text{subject to}
$$

$$
\tilde{x}(\delta^+(v)) - \tilde{x}(\delta^-(v)) = b_v + l(\delta^-(v)) - l(\delta^+(v)) \qquad \text{for } v \in V
$$

$$
0 \leq \tilde{x}_e \leq u_e - l_e \quad \text{for } e \in E
$$

$$
x \ \text{integer valued .}
$$

Note that $\sum_{e \in E} c_e \cdot l_e$ is a constant term as well as $\bar{b}_v := b_v + l(\delta^-(v)) - l(\delta^+(v))$ and thus we have constructed an equivalent MCFP with lower bounds zero. The optimal solution x for the original problem can easily be obtained from the optimal solution $\tilde{x}$ via the transformation $x_e = \tilde{x}_e + l_e$ for $e \in E$.

MCFP and CP as the two standard (general) network flow problems are treated in chapter 7, where we give a short review over different algorithmic approaches. Thereby we will assume furtheron that MCFP is given with zero lower bounds. For some discussions we will also assume that the underlying graph has a single source and a single sink. Now we continue our list with specializations of MCFP.

(N4) The capacitated Hitchcock transportation problem (CHTP)

This is the special case of a MCFP where G is a bipartite digraph with bipartition S and T such that all edges are directed from S to T and $b_s < 0$ for all $s \in S$, $b_t > 0$ for all $t \in T$.

(N5) The transhipment problem (TP)

This is the special case of a MCFP without capacity constraints on the edges, i.e. $u_e = \infty$ for all $e \in E$.

(N6) The Hitchcock transportation problem (HTP)

This is the special case of a TP where G is a bipartite digraph with bipartition S and T such that all edges are directed from S to T and $b_s < 0$ for all $s \in S$ and $b_t > 0$ for all $t \in T$. (Hence we can interpret HTP also as a special case of CHTP with $u_e = \infty$ for all $e \in E$).

The above problems and especially HTP can be designated as the Operations Research Problem(s) at all. The history of this type of problems goes back to the 18th century when the French Academy of Sciences discussed a related civil engineering problem. Monge [1781] may be viewed as the first reference to an OR–article (see also Apell [1928]). In its present form HTP is due to Kantorovich [1942], Hitchcock [1941] and Koopmans [1947]. An early reference to

44

an application of CHTP is Kantorovich and Gavurin [1949]. The first computer code for solving HTP was already developed in 1952 at the National Bureau of Standards and it was capable of solving problems with up to 600 nodes. In 1956 more than 50 % of the linear programming applications in industry were using the HTP–model (cf. Smith [1956]).

Dantzig [1955] showed how CHTP can be reduced to a HTP. Orden [1956] gave a scheme to reduce MCFP into CHTP while Wagner [1959] gave a direct transformation from MCFP to HTP. We will further discuss HTP in a later section.

(N7) The assignment problem (AP)

This is the special case of a HTP in which $|S| = |T|$ and $b_s = -1$ for $s \in S$ and $b_t = 1$ for $t \in T$.

We will analyse (AP) extensively lateron and therefore refer for references on the history of this problem to chapter 10.

Next we introduce two network flow problems which play a fundamental role in the algorithmic treatment of the entire class.

(N8) The shortest path problem (SPP)

Here we have to distinguish two problems. If we want to find the shortest path from a specific node s to all other nodes $v \in V \backslash \{s\}$, then we can think of SPP as a special case of a TP where $b_s = 1 - |V|$ and $b_v = 1$ for $v \in V$, $v \neq s$. If we are interested in the shortest path from s to another specified node t we would define

$$
b_v := \begin{cases} -1 & \text{if } v = s \\ 1 & \text{if } v = t \\ 0 & \text{if } v \in V \backslash \{s, t\} \end{cases}
$$

(N9) The max s–t–flow problem (MFP)

This is the problem

$$max \quad z \quad \text{subject to}$$

$$x(\delta^+(v)) - x(\delta^-(v)) = \begin{cases} -z & \text{for } v = s \\ 0 & \text{for } v \in V \backslash \{s, t\} \\ z & \text{for } v = t \end{cases}$$

$$0 \leq x_e \leq u_e \qquad \text{for } e \in E$$

$$x \quad \text{integer valued ,}$$

where s and t are two prespecified nodes. MFP can easily be cast into an equivalent CP by introducing a "return edge" (t, s) .

Because of their special importance we discuss (MFP) and (SPP) in form of a presection or prologue to network flow problems.

Finally we want to mention

(N10) The maximum cardinality assignment problem (CAP)

Given a bipartite digraph $G = (V, E)$ with bipartition S and T and all edges directed from S to T this is the problem

$$max \quad \sum_{e \in E} x_e \quad \text{subject to}$$

$$x(\delta^+(v)) \leq 1 \quad \text{for } v \in T$$

$$x(\delta^-(v)) \leq 1 \quad \text{for } v \in S$$

$$x_e \in \{0, 1\} \quad \text{for } e \in E$$

which is essentially a max–flow–problem on G and which can be transformed into an AP with $(0, 1)-$ valued edge–weights.

So far we have introduced special cases of the network flow problem. In the following we will analyse the class of matching problems in the same manner.

5.2. Some Special Graph Matching Problems

Here we assume $G = (V, E)$ to be a (undirected) graph and

$c : E \rightarrow \mathbb{R}_+$ a cost function on the set of edges,

$l : E \rightarrow \mathbb{N}_0$ lower bounds on the set of edges,

$u : E \rightarrow \mathbb{N} \cup \{\infty\}$ upper bounds on the set of edges,

$b : V \rightarrow \mathbb{N}$ a capacity function on the set of nodes

(M1) The capacitated b–matching problem (CbMP)

This problem has been introduced already in section 4.2., yet for reasons of completeness we give the formulation here again

$$min \sum_{e \in E} c_e \cdot x_e \quad \text{subject to}$$

$$x(\delta(v)) = b_v \quad \text{for} \ \ v \in V$$

$$l_e \leq x_e \leq u_e \quad \text{for} \ \ e \in E$$

$$x \ \text{integer valued} .$$

As for MCFP we can assume $l_e = 0$ for $e \in E$ since every CbMP with nonzero capacities can in an analogous way be transformed into an equivalent CbMP with lower bounds zero and modified node–capacities $\tilde{b}_v = b_v - l(\delta(v))$. In the case where $\tilde{b}_v < 0$ for a node $v \in V$ we can immediately conclude that CbMP has no feasible solution.

(M2) The b–matching problem (bMP)

This is the special case of a CbMP with

$$u_e = \infty \quad \text{for all} \ \ e \in E \ ,$$

i.e. no capacity constraints on the edges.

The standard reference for bMP — and still the only comprehensive work on this topic — is Pulleyblank [1973].

(M3) The degree–constraint–subgraph problem (DCSP)

This is the special case of a CbMP where

$$u_e = 1 \quad \text{for all} \ \ e \in E \ ,$$

i.e. the requirement $x_e \in \{0,1\}$ for $e \in E$.

With the $\{0,1\}$–condition, bMP becomes the problem to find a subgraph with prescribed degree at any node (and minimal sum over the cost of the edges). DCSP and variants of it have been studied extensively in Urquhart [1967].

(M4) The capacitated Hitchcock transportation problem (CHTP)

and

(M5) The Hitchcock transportation problem (HTP)

can be viewed as specializations of CbMP resp. bMP with the underlying graph $G = (V, E)$ a bipartite graph.

48

(M6) The 1–matching problem (1MP)

This is the bMP with $b_v = 1$ for all $v \in V$ or equivalently formulated

$$min \sum_{e \in E} c_e \cdot x_e \quad \text{subject to}$$

$$x(\delta(v)) = 1 \quad \text{for} \quad v \in V$$

$$x_e \in \{0,1\} \quad \text{for} \quad e \in E \, .$$

Here we have to find a collection of edges (subgraph) such that every node is incident with exactly one edge out of this collection (and the sum over the edge weights associated with edges from the collection has to be minimal). Structures or subgraphs having this property are called *1–factors* or *perfect 1–matchings*.

As will be shown, 1MP is the key problem within the entire class of (general) matching problems — it "contains" already the whole complexity of the entire class as well as the genuine charme. For that reason we will simply call this problem matching problem — if no confusion is to be expected — and abbreviate 1MP to MP.

The problem is often called *1–factor problem*, too, and is a celebrated object in Graph Theory (cf. Tutte [1947], Berge [1957]). From an algorithmic point of view MP is immediately connected with the work of Jack Edmonds [1965], who's ideas, first applied to tackle MP, opened up a new scope of vision within the field of combinatorial programming.

(M7) The assignment problem (AP)

can be viewed as a special case of 1MP where the underlying graph $G = (V, E)$ is bipartite.

(M8) The cardinality 1–matching problem (CMP)

This is the problem

$$max \sum_{e \in E} x_e \quad \text{subject to}$$

$$x(\delta(v)) \leq 1 \quad \text{for} \quad v \in V$$

$$x_e \in \{0,1\} \quad \text{for} \quad e \in E .$$

which can be transformed into a 1MP with $(0, 1)$–valued edge–weights.

Finally we mention again

(M9) The maximum cardinality assignment problem (CAP)

which can be viewed as specialization of CMP for bipartite graphs.

The matching and network problems listed above are all celebrated objects of research in Operations Research and Graph Theory. The list has already demonstrated the close connection between network flow and graph matching since some standard problems — AP and HTP for instance — can be viewed and treated as special network flow as well as special matching problems. Thus in addition to the common generalization - the general matching problem in a bidirected graph - those problems built another interface between the two classes and they will be of special interest with respect to algorithmic investigations.

The next figure reflects the hierarchy among these problems again.

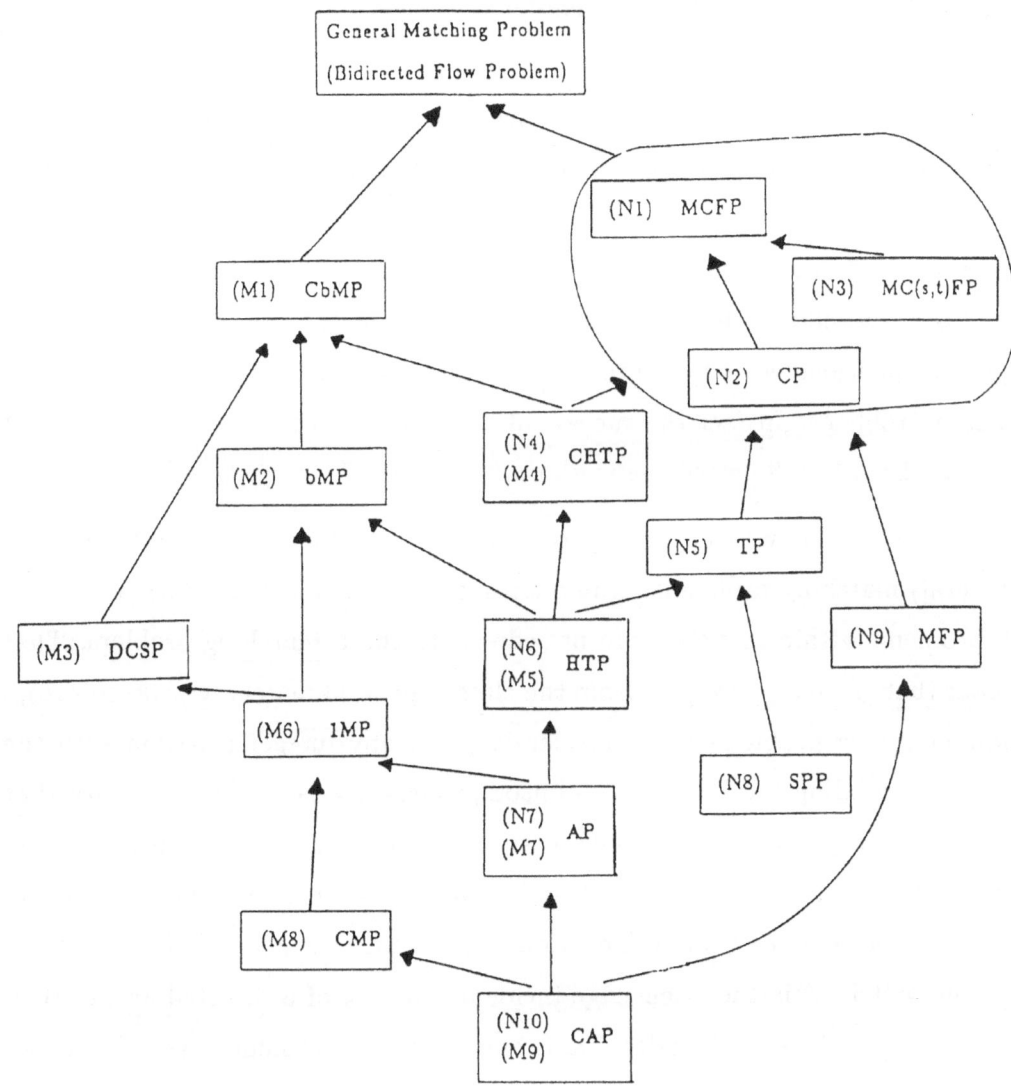

Figure 5.1. Hierarchy among matching problems

($\overline{|A|} \leftarrow \overline{|B|}$ stands for "B is special case of A")

51

5.3. Reductions for Matching Problems

In the last section we have introduced a quite comprehensive list of special matching and network flow problems and established a hierarchy among these problems. All these special subproblems have found considerable interest in Operations Research and Graph Theory and special algorithmic approaches have been develloped for all of them. Thereby one can observe the common "process of learning" where procedures are either *specialized* or *extended* to related problems. In fact the success of network flow programming techniques has significantly influenced the work on "matchings".

In this section we will present some reduction techniques within the class of (general) matching problems. The main result thereby is the information that all problems within this class can be reduced to the 1–matching problem. That means that given any problem from the class of general matching problems resp. bidirected flow problems (or equivalently given any integer program with the matrix A fulfilling the Edmonds–Johnson property) we can associate with that problem a graph $G = (V, E)$ with cost function $c : E \rightarrow \mathbb{R}_+$ such that any min–cost–1–matching in G induces an optimal solution of the original problem. Moreover we will see that if the original problem is a network flow problem i.e. the matrix A is the node–edge incidence matrix of a directed graph, then $G = (V, E)$ will be a bipartite graph and thus the problem is reduced to an assignment problem.

The possibility of reducing any network flow problem to an assignment problem or bipartite 1–matching problem serves to prove the diction that "network programming is bipartite programming" and for Lawler [1976] this is the basis upon which the *theory of bipartite matching* can be considered to be coextensive with the *theory of network flows*. On the same basis we can therefore consider *1–matching theory* to be coextensive with the *theory of programming in bidirected graphs* or the *theory of bidirected network flows*.

Although any algorithmic principle for a problem in this class can be synthesized from 1-matching solution techniques and crystalized to its essential combinatorial ingredients when specialized to 1MP, these results are not equivalently valuable from an algorithmic point of view. The process of "problem reduction" and "adaption of algorithms" does not always lead to a "good" algorithm for the original problem. Thus we are interested in reductions which preserve the property of efficiency; such reductions are called *polynomial reductions*.

We want to state this relation more formal. Assume two problems P_1 and P_2 – bMP and 1MP for instance. Then a mapping $R : P_1 \to P_2$ which assigns to every instance I_1 in P_1 an instance $I_2 = R(I_1)$ in P_2 is called reduction of P_1 to P_2 if the optimal solutions (optimal values) of I_1 and I_2 are identical, thus I_1 can be solved by solving I_2 instead. The reduction $R : P_1 \to P_2$ is called *polynomial* if size $(R(I_1))$ is polynomial in size (I_1), i.e. given an algorithm for solving P_2 which is polynomial in the input size of the instance of P_2, problem P_1 can be solved in polynomial time by means of the reduction R and the algorithm for solving P_2 .

Thus we are especially interested in identifying polynomial reductions. With respect to this concept we will extract two key problems

- the 1-matching problem and

- the b-matching problem,

where this list has to be extended by

- the assignment problem and

- the transportation problem

if we want to point out the additional property of bipartiteness when reducing network flow problems.

(i) Reduction of DCSP to 1MP

The first classical reduction is due to Berge [1962]. For any node $v \in V$ we define $\Delta_v := d(v) - b_v$, the difference between the degree of v in G and the capacity of v i.e. the prescribed degree of v in the subgraph to be constructed. Now v is replaced by its "substitute" S_v the complete bipartite graph $K_{\Delta_v, d(v)}$ as the following figure demonstrates

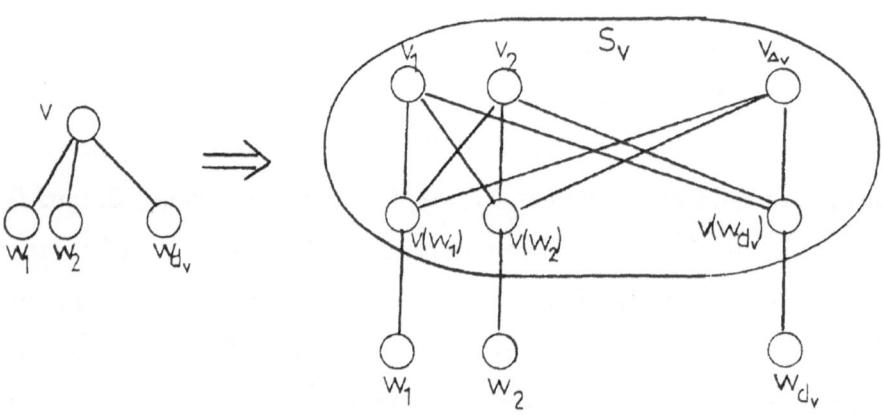

Figure 5.2. Substitution of v by $K_{\Delta_v, d(v)}$

The graph obtained from G by substituting every node v by its appropriate substitute is denoted by $\tilde{G} = (\tilde{V}, \tilde{E})$.

Now given a feasible solution $x : E \to \{0,1\}$ for the DCSP on G we obtain a solution $\tilde{x} : \tilde{E} \to \{0,1\}$ which is feasible for the 1MP on $\tilde{G}$ in the following way.

First set

$$\tilde{x}_{v(w),w(v)} := \begin{cases} 1 & \text{if } x_{vw} = 1 \\ 0 & \text{else} \end{cases}$$

and then match the remaining $2\Delta_v$ nodes in every substitute S_v pairwisely to complete the definiton of $\tilde{x}$.

Moreover if we define

$$\tilde{c}_{v(w),w(v)} = c_{vw} \quad \text{for all } \{v,w\} \in E \quad \text{and}$$

$$\tilde{c}_{v_i,v(w_j)} = 0 \quad \text{for } \{v_i, v(w_j)\} \quad \text{an edge in the substitute of } v$$

then $\sum_{e \in \tilde{E}} \tilde{c}_e \cdot \tilde{x}_e = \sum_{e \in E} c_e \cdot x_e$, hence the objective function values are equal.

On the other hand let $\tilde{x}$ be a perfect matching in $\tilde{G}$ then for every substitute S_v we have

$$\tilde{x}(\delta(S_v)) = b_v$$

and the following mapping $x := E \to \{0,1\}^E$ is a solution for the DCSP on G

$$x_{vw} := \begin{cases} 1 & \text{if } \tilde{x}_{v(w),w(v)} = 1 \\ 0 & \text{else} \end{cases}$$

Moreover $c'x = \tilde{c}'\tilde{x}$ holds again.

By the above transformation DCSP can be solved by any 1MP–algorithm and obviously this transformation is polynomial, too.

Since

$$\left|\tilde{V}\right| = \sum_{v \in V}(\Delta_v + d_v) \le 4\,|E| \quad \text{and}$$

$$\left|\tilde{E}\right| = |E| + \sum_{v \in V}\Delta_v \cdot d_v \le |E| + 2\,|V|\,|E| \ ,$$

applying an algorithm which is polynomial in $\left|\tilde{V}\right|$ and $\left|\tilde{E}\right|$ to solve the 1MP on $\tilde{G}$ leads to an algorithm for solving DCSP which is polynomial in $|V|$ and $|E|$, the size of the DCSP.

Recently, Gabow [1983] has analysed the application of the 1MP–algorithm to the graph $\tilde{G}$ and based on some fundamental observations he could show that a much "sparser" substitution is sufficient for reducing DCSP to 1MP.

(ii) Reduction bMP $\to$ 1MP

It is obvious that bMP can be transformed into a DCSP on a *multigraph* by replacing every edge $e = \{u,v\}$ by $k = min\,\{b_u, b_v\}$ copies. Thereafter we could reduce the corresponding DCSP to 1MP. In the following we give a direct reduction which is also of interest in connection with establishing the b–matching polytope (cf. Araoz et. al [1983]).

The new graph $\tilde{G} = (\tilde{V}, \tilde{E})$ obtained from G is given by

$$\tilde{V} := \bigcup_{v \in V} \{v_1, \ldots, v_{b_v}\}$$

$$\tilde{E} := \{\{v_i, w_j\} \mid \{v, w\} \in E \ , \ v_i, \ w_j \in \tilde{V}\}$$

and the cost function $\tilde{c} : \tilde{E} \to \mathbb{R}_+$ is defined by

$$\tilde{c}_{v_i w_j} := c_{vw} \qquad \text{for } \{v_i, w_j\} \in \tilde{E} \ .$$

Thus we have replaced each vertex v by b_v copies and we have introduced $b_v \cdot b_w$ copies of each edge $\{v, w\} \in E$. Now let x be a b–matching in G, then we construct an associated 1–matching $\tilde{x}$ in $\tilde{G}$ with $c'x = \tilde{c}'\tilde{x}$ as follows:
Pick any edge $e = \{v, w\} \in E$ with $x_e > 0$ and set

$$\tilde{x}_{v_i w_j} := \begin{cases} 1 & \text{for } 1 \le i = j \le x_e \\ 0 & \text{else} \end{cases}$$

and delete from $\tilde{G}$ all the vertices $v_1, \ldots, v_{x_e}, w_1, \ldots, w_{x_e}$ together with their incident edges (on these edge we set $\tilde{x}$ to zero). Now repeating this process for all $e = \{v, w\}$ with $x_e > 0$ — after appropriately renumbering the remaining copies of nodes which have already been involved — yields the desired 1–matching $\tilde{x}$ in $\tilde{G}$ with $\tilde{c}'\tilde{x} = c'x$.
Now suppose a 1–matching $\tilde{x}$ in $\tilde{G}$ is given. Then we define an associated b–matching $x \in \mathbb{N}_0^E$ in the following way. Set

$$x_e := \sum_{i=1}^{b_v} \sum_{j=1}^{b_w} \tilde{x}_{v_i w_j} \qquad \text{for } e = \{v, w\} \in E \ .$$

Then $x(\delta(v)) = b_v$ for all $v \in V$, thus x is a b–matching in G and more obviously $\tilde{c}'\tilde{x} = c'x$ holds, too.

The above transformation does <u>not</u> lead to a polynomial algorithm for bMP. Assume we have a polynomial algorithm for solving the 1MP on $\tilde{G}$ of complexity

$O(|V|^\alpha |E|^\beta)$. Since $\left|\bar{V}\right| = \sum_{v \in V} b_v (=: b)$ we obtain an algorithm for solving bMP which is polynomial in b yet not in $\log b$, the input size of the "node capacities".

<u>(iii) Reduction CbMP → bMP</u>

The reduction which we present here occurs at least as early as Edmonds [1967] and it is influenced by similar reductions for network flow problems. We may assume $u_e > 0$ for all $e \in E$ since otherwise we could delete the edge e . Now we replace every edge $e = \{v, w\}$ by three edges

$$e = \{v, p_e\} \, , \ e_1 = \{p_e, q_e\} \quad \text{and} \quad e_2 = \{q_e, w\}$$

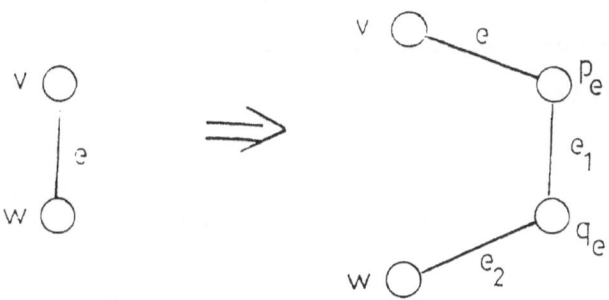

<u>Figure 5.3.</u> Transformation of edge $e = \{v, w\}$

Thus we obtain a new graph $\tilde{G} = (\bar{V}, \bar{E})$ with

$$\bar{V} := V \cup \{p_e \mid e \in E\} \cup \{q_e \mid e \in E\} \quad and$$
$$\bar{E} := \bigcup_{e \in E} \{e, e_1, e_2\}$$

For the new nodes we set

$$b_{p_e} := b_{q_e} := u_e$$

and for the new edges we define

$$c_{e_1} := c_{e_2} := 0 \ .$$

Now for any b–matching $\bar{x}$ in $\tilde{G}$ we have

$$\bar{x}_e + \bar{x}_{e_1} = b_{p_e} = b_{q_e} = \bar{x}_{e_1} + \bar{x}_{e_2}$$

which implies $\tilde{x}_e = \tilde{x}_{e_2} \le u_e$. Hence every b–matching $\tilde{x}$ in $\tilde{G}$ induces a b–matching x in G fulfilling the capacity condition by setting $x_e = \tilde{x}_e$, moreover $c'x = \tilde{c}'\tilde{x}$ holds.

Obviously this transformation is polynomial, too, and, if G is bipartite, then $\tilde{G}$ is bipartite, too.

(iv) Reduction GMP → CbMP

This transformation is also due to Edmonds [1967] . Every node $v \in V$ is replaced by two nodes v^+ and v^- which are joined by a new (undirected) edge $\{v^+, v^-\}$. Now all edges from $\delta^+(v)$ are introduced as undirected edges incident with v^- , and all edges in $\delta^-(v)$ are identified with undirected edges incident with v^- with the costs and capacities of these edges unchanged.
In addition we define

$$\tilde{c}_{v^+v^-} := 0$$

$$\tilde{u}_{v^+v^-} := \infty \quad \text{for all } v \in V$$

and $\tilde{b}_{v^+}$ and $\tilde{b}_{v^-}$ appropriately large such that

$$\tilde{b}_{v^+} - \tilde{b}_{v^-} = b_v \quad \text{for all } v \in V .$$

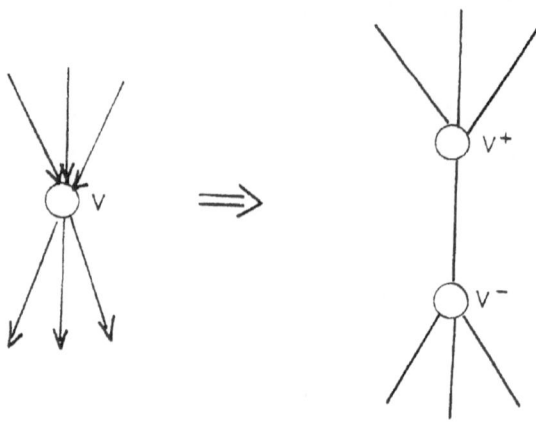

Figure 5.4. Transformation from G to $\tilde{G}$

Now the capacitated b–matching problem on $\tilde{G}$ and the general matching problem on the bidirected graph G are equivalent.

In fact, given a feasible solution x for the GMP on G we define

$$\tilde{x}_e = x_e \quad \text{for} \ e \in E \qquad \text{and}$$

$$\tilde{x}_{v^+v^-} = b_{v^+} - x(\delta_G^+(v)) \ .$$

With this definition $\tilde{c}'\tilde{x} = c'x$ holds and the capacity conditions for any v^+ in $\tilde{G}$ and the capacity conditions for all edges in $\tilde{G}$ are fulfilled.

Moreover the following relation holds for all v^- in $\tilde{G}$:

$$\tilde{x}(\delta_{\tilde{G}}(v^-)) = x(\delta_G^-(v)) + \tilde{x}_{v^+v^-} = x(\delta_G^-(v)) - x(\delta_G^+(v)) + b_{v^+}$$

$$= -b_v + b_{v^+} = b_{v^-}$$

i.e. the capacity conditions for all v^- in $\tilde{G}$ are fulfilled, too. The nonnegativity of $\tilde{x}$ is assured if b_{v^+} and b_{v^-} are chosen appropriately large. Hence $\tilde{x}$ is feasible for the CbMP on $\tilde{G}$.

On the other hand given a solution $\tilde{x}$ for the CbMP on $\tilde{G}$ then the restriction $x := \tilde{x} \mid_E$ of $\tilde{x}$ to the edges in E is a feasible solution for the GMP on G and obviously $\tilde{c}'\tilde{x} = c'x$ holds, too.

The above reduction is polynomial since $\left|\tilde{V}\right| = 2\,|V|$, and $\left|\tilde{E}\right| = |E| + |V|$ and it is easy to see that if the general matching problem is defined on a directed graph G then the graph $\tilde{G}$ will be bipartite.

Using the reductions (iv), (iii) and (ii) any GMP on a bidirected graph can be reduced to a 1MP. Hence we can consider the 1–matching problem as the key problem within this class. Also we have shown this way that any MCFP can be reduced to an assignment problem. Only the last step in this chain of reductions is not polynomial . Thus if we want to show polynomiality of all the problems within this class, we have to discuss bMP and HTP as well.

We have shown the reductions for the "flow problems" as a byproduct of the reductions for the "matching problems". Orden [1956] has deviced a method which transforms a MCFP into a CHTP and Dantzig [1963] has given a reduction of CHTP to HTP. In fact it were these transformations which inspired the more general reductions for bidirected graphs and matching problems. Finally we should mention that Wagner [1959] has given a direct reduction for MCFP into HTP.

In the following we present another reduction which illustrates another connection between "network flow" and "matching".

(v) Reduction bMP with $b \equiv 0 \mod 2 \rightarrow$ HTP

Here every node $v \in V$ will be replaced by two nodes v_1 and v_2 and the capacities are split into $b_{v_1} := b_{v_2} := \frac{1}{2} b_v$. For any edge $\{v, w\}$ we introduce two edges (v_1, w_2) and (w_1, v_2) with the same cost as the original edge.
Now let $\tilde{x}$ be a solution for the HTP on the new graph $\tilde{G}$, then we obtain a solution x for the original bMP having the same total cost by setting

$$x_{vw} = \tilde{x}_{v_1 w_2} + \tilde{x}_{w_1 v_2} .$$

On the other hand every solution x of the original bMP induces a solution $\tilde{x}$ for the associated HTP of the same total cost setting.

$$x_{v_1 w_2} := x_{w_1 v_2} := \frac{1}{2} x_{vw} .$$

For the reduction we have used the "matching–formulation" of HTP. Yet it is obvious that we could have used as well the "network–flow–formulation" of HTP using a directed bipartite graph.

60

PART III

NETWORK FLOW ALGORITHMS REVISITED

Numerous computational and theoretical works have developed a variety of algorithms for solving network flow problems. Often known methods were reinvented by different authors when approaching from different points of view or when generalizing or specializing existing methods. Jensen and Barnes [1980] have classified network flow procedures into four classes

- primal methods

- primal node–infeasible methods

- dual edge–infeasible methods

- primal–dual methods

and they describe implementations of algorithms out of these classes. Yet the above classification is not standard and different other schemes have been proposed and used. Thereby it turned out that several algorithms could not be classified uniquely, they could be interpreted in different ways which led to different assignments to classes.

Recently, Hassin [1983] has divided the most commonly used network flow algorithms into only two classes

- primal methods and

- dual methods,

showing the unifying combinatorial background behind algorithms out of the same class.

In the literature on network optimization problems there is often an equivalence or near equivalence between the primal–dual and the out–of–kilter network algorithm assumed (cf. Dantzig [1963], Simmonard [1966]).

In a pioneering work Zadeh [1980] has further analysed the combinatorial backbone of different network algorithms and he discovered a certain "partial equivalence" among all commonly used approaches. In fact he could show that these procedures perform the same sequence of "steps" if they are started from appropriate starting points (initial flow and dual (node) weights) and ties are broken in the same manner. This result indicates that there is a certain "natural" strategy inherent in all (efficient) solution strategies and the computational advantages of specific algorithms may lie in the fact that they are more *flexible* with respect to starting conditions etc. We will come back to this more philosophical question concerning the combinatorial backbone of network flow algorithms later on.

In this part we will analyse and approach the network flow problem from different points of view and thereby we motivate and develop some of the most common algorithms and we will demonstrate their near equivalence.

From the contributions in literature it is evident that the algorithmic treatment of matching problems has benefited substantially from results and algorithmic principles developed for network flow problems. Under the algorithmic aspect "matching theory" was always about ten years behind "flow theory". Now the purpose of this section's analysation of network flow problems and algorithms is to guide the analog analysation of the combinatorially more complex (real) matching problems.

Chapter 6. Prologue: Two Apparently "Easier" Network Flow Problems

In the following we shortly discuss basic algorithmic approaches to two fundamental network flow problems

- the shortest path problem (SPP) and

- the max–flow problem (MFP)

which are major analytical components of numerous complex quantitative transportation and communication models.

We have already introduced both problems as special cases of MCFP. In the hierarchy of network flow problems both problems are somewhat "aside" the main stream of reduction at a lowest level. This indicates that SPP and MFP are in a sense "easier" than other network flow problems.
In fact shortest–path computations and max–flow computations are essential ingredients of approaches for solving general MCFP's. On this basis we focus on the procedural aspects of SPP and MFP which are of importance for MCFP.

There are a number of interesting results known concerning the theoretical background of SPP–labeling techniques resp. MFP–labeling techniques. (cf. for instance Papadimitriou and Steiglitz [1982] where both, the well–known Dijkstra–labeling method and the Ford–Fulkerson labeling method, are developed as specializations of the general primal–dual method for solving linear programs) But we do not discuss these more theoretical aspects of SPP– and MFP–algorithms here.

6.1. The Shortest Path Problem

Given a digraph $G = (V, E)$ suppose each edge $(i, j) \in E$ is assigned a numerical length $c_{ij} \in \mathbb{R}$, then we define the length $l(P)$ of a directed path P as the sum over all the edge–lengths of edges contained in P . We can distinguish three different shortest path problems (SPP) in G :

The single pair problem

Find a shortest path from a given "source" s to a given "sink" t , a so–called shortest s–t–path.

The single source problem

Find shortest paths from a given "source" s to all other nodes in V .

The all–pairs problem

Find shortest paths between all possible pairs (s, t) .

In the following we will discuss the single–source problem only since

- the all–pairs problem can be interpreted and solved as a sequence of single–source problems,

- the methods which we present can also be used for solving the single–pair problem.

In the following we will use $n := |V|$ to measure the size of a SPP. From an algorithmic point of view three fundamental cases must be considered:

(i)   all c_{ij}'s are nonnegative,

(ii)   some c_{ij}'s are negative, but G does not contain any cycle of negative length, a so–called negative cycle,

(iii) G contains negative cycles.

In the latter case it may not be possible to "solve" the shortest path problem:

Theorem 6.1

There is a shortest path from s to t iff

(i)   t is reachable from s and

(ii)   no path from s to t contains a cycle of negative length.

Thus the algorithms which we present here will either output a shortest s–t–path or discover a negative cycle in which case they also halt.

The problem of discovering negative cycles is an important subproblem in some approaches for MCFP. Hence shortest path techniques can be used to solve these subproblems.

A spanning tree rooted at s is called *(shortest) path tree* if each path from s to $v \in V \backslash \{s\}$ in the tree is a (shortest) s–v–path.

Theorem 6.2

G contains (shortest) paths from s to all other nodes in V iff G contains a (shortest) path tree rooted at s .

Thus the single source problem can be solved by constructing a shortest path tree (and the algorithms for solving the single pair problem will at least partially built up such a shortest path tree).

For $v \in V$ let us define $l(v)$ the length of a shortest path from s to v , then the following basic theorem holds.

Theorem 6.3   (Characterization of shortest path trees)

T is a shortest path tree iff for every edge $(i, j) \in E$

$$l(i) + c_{ij} \geq l(j)$$

where $l(k)$ is the length of the path from s to k in T .

The above theorem motivates the following general procedure which is ascribed to Moore [1957] :

<u>Procedure 6.1.</u>   **General label correcting method for solving SPP**

<u>Step 0:</u> Set $l(s) := 0$

$l(v) := \infty$ for $v \in V \setminus \{s\}$

$p(v) := 0$ for $v \in V$

<u>Step 1:</u> Determine edge $(u, v) \in E$ with $l(u) + c_{uv} < l(v)$

if no such edge exists, Stop: a shortest path tree rooted

at s has been constructed

otherwise goto <u>Step 2</u>

<u>Step 2:</u> Set $p(v) := u$

$l(v) := l(u) + c_{uv}$

and goto <u>Step 1</u>

(Note: The shortest path from s to a node $v \in V$ can be traced back from v using successively the "predecessor–label" $p(\ )$.)

The following variant is attributed to Bellman [1958] and Ford [1956] . Here the edges of the tree are investigated in a prescribed ("natural") order:

Initially we set $l^0(s) := 0$ and $l^0(v) := \infty$ for $v \in V \setminus \{s\}$ as above and then we compute the "$(k + 1)$st order approximation" from the kth order approximation as follows:

$$l^{k+1}(u) = min\{l^k(u), min\{l^k(v) + c_{vu} \mid (v, u) \in E\}\} .$$

If the network does not contain negative cycles then for all nodes $v \in V$

$$l(v) := l^{n-1}(v)$$

is the length of a shortest s–v–path, thus (at most) $n - 1$ iterations are necessary. (The computation may be stopped whenever $l^{k+1}(v) = l^k(v)$ for <u>all</u> $v \in V$). Each iteration requires $O(n^2)$ elementary steps. Hence the computational complexity of the Bellman–Ford procedure is $O(n^3)$.

The Bellman–Ford method can also be used to detect negative cycles in G :

Theorem 6.4

Let the directed graph G have a path from s to every other node. Then G containes a negative cycle iff

$$l^n(v) < l^{n-1}(v)$$

for at least one $v \in V$.

Thus we have to carry out the Bellman–Ford method for one additional iteration to check the existence of negative cycles.

A second type of method which works only in the case of nonnegative $c'_{ij}s$ is the following

---

Procedure 6.2.   **General label setting method for solving SPP**

Step 0: Set $l(s) := 0$

$\quad\quad l(v) := \infty$ for $v \in V \backslash \{s\}$

$\quad\quad p(v) := 0$ for $v \in V$

$\quad\quad S := \{s\}$

Step 1: Determine $q \in S$ and $u \in V \backslash S$ s.t.

$\quad\quad l(q) + c_{qu} = min\{l(v) + c_{vw} \mid (v,w) \in \delta^-(S)\}$

$\quad\quad$ if no such pair exists, Stop: a shortest path tree rooted

$\quad\quad\quad\quad$ at s has been constructed

$\quad\quad$ otherwise goto Step 2

Step 2 Set $l(u) := l(q) + c_{q,u}$

$\quad\quad p(u) := q$

$\quad\quad S := S \cup \{u\}$

$\quad\quad$ and goto Step 1.

The basic label–setting algorithm is credited to Minty [1957]. Dijkstra [1959] has developed a highly efficient variant in which for every node $w \in V \setminus S$ the value

$$l(w) = min\{l(v) + c_{vw} \mid (v,w) \in \delta^-(S)\}$$

is stored together with the associated node $v \in S(=: p(w))$.

Then the minimum in Step 1 is not taken over edges in $\delta^-(S)$ anymore but over the nodes in $\bar{S} := V \setminus S$:

Procedure 6.3.   Dijkstra's labeling method for solving SPP

Step 0: Set $l(s) := 0$

$$l(v) := \begin{cases} c_{sv} & \text{for} \quad (s,v) \in E \\ \infty & \text{otherwise} \end{cases}$$

$$p(v) := s \quad \text{if} \quad l(v) < \infty$$

$$S := \{s\}$$

Step 1: Determine $w \in \bar{S}$ with

$l(w) = min\{l(v) \mid v \in \bar{S}\}$

If $l(w) = \infty$ or $\bar{S} = \emptyset$, Stop: a shortest path tree

rooted at s has been constructed

otherwise goto Step 2.

Step 2: Set $S := S \cup \{w\}$

$l(v) := min\{l(v), l(w) + c_{wv}\}$ for $v \in \bar{S}$

$p(v) := w$ if $l(v)$ has changed

goto Step 1

Dijkstra's method requires $(n-1)(n-2)$ comparisons and $(n-1)(n-2)/2$ additions overall, thus the method is $O(n^2)$ in complexity.

Various efficient data–structures have been developed for implementing Moore's method and Dijkstra's method (cf. Pape [1974], Dial [1969] and others). A comprehensive computational analysis of alternative SPP–algorithms has been published by Gilsinn and Witzgall [1973], Dial et al. [1979] and Gallo et al. [1982].

It is beyond the scope of this study to even only outline the main computational results presented in literature. None of the common algorithms turned out to be the overall winner. The behaviour of the different procedures has shown to be highly dependent on the structure of the graph (density, grid–structure etc.) and the range of the $c'_{ij}s$. Yet the message of these computational studies is clear: An optimal combination of the basic algorithm and the data–structures developed for SPP–procedures depending on the graph structure and cost–range seems to be the key–issue when solving a given SPP.

Recently several surveys on SPP-methods have been published where new classifications of SPP–algorithms are suggested and certain "prototype methods" are given with the property that all the common SPP–approaches can be interpreted as special implementations of this prototype, cf. Gallo and Pallottino [1983], Glover et al. [1982].

6.2. The Max–Flow Problem

Let $G = (V, E)$ be a directed graph with two distinguished nodes, a *source* s and a *sink* t, and a positive capacity u_e on every edge $e \in E$.

A *s–t–flow* on G is a mapping $x : E \to \mathbb{R}_+ \cup \{0\}$ s.t.

$$x(\delta^+(v)) - x(\delta^-(v)) = 0 \quad \text{for all } v \in V \backslash \{s, t\} .$$

The *value* $z := z(x)$ of a flow x is the net–flow out of the source, i.e. $z = x(\delta^-(s)) - x(\delta^+(s))$.

A flow is called *feasible* if $0 \leq x_e \leq u_e$ for all $e \in E$ holds, and the *max–flow–problem* (MFP) is then to find a feasible flow of maximal value. This problem can be formulated as linear program

$$max \quad z \tag{MFP}$$

$$x(\delta^+(v)) - x(\delta^-(v)) = \begin{cases} -z & \text{for } v = s \\ 0 & \text{for } v \in V \backslash \{s, t\} \\ z & \text{for } v = t \end{cases}$$

$$0 \leq x_e \leq u_e \quad \text{for } e \in E .$$

W.l.o.g. we may assume $\delta^+(s) = \delta^-(t) = \emptyset$, i.e. $z(x) = x(\delta^-(s))$.
Then the dual program associated with (MFP) is given by

$$min \sum_{e \in E} w_e \cdot u_e \quad \text{subject to} \tag{DMFP}$$

$$- y_{t(e)} + y_{h(e)} + w_e \geq 0 \quad \text{for } e \in E$$

$$y_s - y_t \geq 1$$

$$w_e \geq 0 \quad \text{for } e \in E$$

$$y_v \quad \text{unrestricted in sign} \quad \text{for } v \in V .$$

Now DMFP can also be interpreted in a combinatorial manner by introducing a concept which in this sense is dual to the concept of flows:

A s-t-cut in G is a partition $(W, \bar{W})$ of the nodes in V into sets W and $\bar{W}$ such that $s \in W$ and $t \in \bar{W}$. The capacity of a s-t-cut is defined as

$$\text{cap } (W, \bar{W}) := u(\delta^-(W)) .$$

The value of any feasible s-t-flow cannot exceed the capacity of any s-t-cut. For suppose x is a feasible flow and $(W, \bar{W})$ is a s-t-cut, then we get

$$z = \sum_{v \in W} [x(\delta^-(v)) - x(\delta^+(v))] = x(\delta^-(W)) - x(\delta^+(W))$$

i.e. the value of any s-t-flow equals the net s-t-flow through any s-t-cut.

But $x_e \geq 0$ implies $x(\delta^+(W)) \geq 0$ and $x_e \leq u_e$ implies $x(\delta^-(W)) \leq u(\delta^-(W))$ hence the following weak duality relation holds:

$$max \ \{z(x) \mid x \text{ feasible } s\text{-}t\text{-flow}\} \leq min \ \{\text{cap } (W, \bar{W}) \mid (W, \bar{W} \ s\text{-}t\text{-cut}\} .$$

Each s-t-cut $(W, \bar{W})$ determines a feasible solution for DMFP with value cap $(W, \bar{W})$ by setting

$$y_v := \begin{cases} 1 & \text{for } v \in W \\ 0 & \text{for } v \in \bar{W} \end{cases}$$
$$w_e := \begin{cases} 1 & \text{for } e \in \delta^-(W) \\ 0 & \text{otherwise.} \end{cases}$$

Hence the validity of the weak duality relation follows immediately from LP-duality. Moreover it can be shown that there is an optimal solution to DMFP which corresponds to a s-t-cut. Hence the following theorem holds:

Theorem 6.5.   *Max–flow–min–cut–theorem   (Ford and Fulkerson [1956])*

The value of the maximum feasible s–t–flow equals the capacity of the minimum s–t–cut. Moreover a flow x and a cut $(W, \bar{W})$ are jointly optimal for MFP and DMFP iff

$$x_e = 0 \quad \text{for } e \in \delta^+(W)$$

$$x_e = u_e \quad \text{for } e \in \delta^-(W) \, .$$

The original proof by Ford and Fulkerson [1956] did not use LP–duality and was based on purely combinatorial arguments. The proof was done in a constructive way by giving a (labeling–) method which produces the optimal s–t–flow and an associated minimal s–t–cut.

Let x be any feasible s–t–flow ($x \equiv 0$ for instance) then we define an associated *incremental graph* $G(x) = (V, E(x))$ in the following way:

Let $E(x) := E^+(x) \cup E^-(x)$ with

$$E^+(x) := \{(i, j) \mid (i, j) \in E \, , \, x_{ij} < u_{ij}\}$$
$$E^-(x) := \{(j, i) \mid (i, j) \in E \, , \, x_{ij} > 0\}$$

and we define new capacities

$$u_{ij}(x) = \begin{cases} x_{ji} & \text{for } (i, j) \in E^-(x) \\ u_{ij} - x_{ij} & \text{for } (i, j) \in E^+(x) \, . \end{cases}$$

The edges in $E^+(x)$ are called *forward edges* and the edges in $E^-(x)$ are called *reverse edges*. A node $w \in V$ is called *reachable* from s in G if there exists a directed path from s to w in $G(x)$.

Now two cases may occur:

<u>Case 1:</u> Sink t is reachable from source s

i.e. there exists a directed (simple) s–t–path $P = (V(P), E(P))$ in $G(x)$.

In that case define $\epsilon(P) := min\{u_{ij}(x) \mid (i,j) \in E(P)\}$ and set

$$\bar{x}_{ij} := \begin{cases} x_{ij} + \epsilon(P) & \text{for } (i,j) \in E(P) \cap E^+(x) \\ x_{ij} - \epsilon(P) & \text{for } (j,i) \in E(P) \cap E^-(x) \\ x_{ij} & \text{otherwise .} \end{cases}$$

Then $\bar{x}$ is a feasible s–t–flow with $z(\bar{x}) = z(x) + \epsilon(P) > z(x)$. A path P in G corresponding to a directed s–t–path in $G(x)$ is called a *flow–augmenting path* and we will abbreviate

$$\bar{x} := x \oplus \epsilon(P) \cdot x(P)$$

and we call this transformation a *flow–augmentation*.

<u>Case 2:</u> Sink t is not reachable from source s

In this case define

$$W := \text{ set of nodes reachable from } s$$

$$\bar{W} := V \backslash W .$$

Then $(W, \bar{W})$ is a s–t–cut and the following relations hold:

$$e \in \delta^-(W) \quad \Rightarrow \quad x_e = u_e$$

$$e \in \delta^+(W) \quad \Rightarrow \quad x_e = 0 .$$

Thus cap $(W, \bar{W}) = z(x)$ and hence $(W, \bar{W})$ and x are both optimal for DMFP and MFP, respectively.

With the above we have also proven two principle theorems of network flow theory:

Theorem 6.6   *(Augmenting path theorem, Ford and Fulkerson [1962])*

A feasible flow is maximal iff it admits no flow–augmenting path.

This theorem motivates the following scheme for solving MFP

<u>Procedure 6.4.</u>   **Augmenting path algorithm for solving MFP**

<u>Step 0:</u> Find an initial feasible s–t–flow x ($x \equiv 0$ for instance)

<u>Step 1:</u> Check for a flow–augmenting path P in $G(x)$.

　　　　If no such path P exists, Stop: x is optimal.

　　　　Otherwise, given a flow–augmenting path P , goto <u>Step 2</u>.

<u>Step 2:</u> Set $x := x \oplus \epsilon(P) \cdot x(P)$ and goto <u>Step 1</u>.

The second fundamental theorem is the following

Theorem 6.7.   *(Integral flow theorem)*
If all capacities are integers then there is a maximal s–t–flow which is integral.

The last theorem follows immediately from the fact that starting from an integral flow ($x \equiv 0$ for instance) every next flow will be integral too, since $\epsilon(P) \in \mathbb{N}$ holds. Moreover if all capacities are integer the above scheme will terminate after at most z^* iterations, where z^* is the maximal flow value. Even if all the capacities are mutually commensurable, i.e. there exists some $D > 0$ such that every u_e is an integral multiple of D , the scheme will produce the optimal solution in a finite number of iterations.

Yet, Ford and Fulkerson [1962] could give an example with capacities which are not commensurable where the above scheme needs not terminate and it even may converge to a non–optimal flow. The problem of finding a flow–augmenting path P and the determination of $\epsilon(P)$ can be handled simultaneously. According to theorem 6.2. a (directed) path connecting s and t in $G(x)$ can be found by constructing a (directed) path tree in $G(x)$ rooted at s . Similar to the SPP–procedures, nodes v contained in the tree receive two labels

- the predecessor label $p(v)$ and

- the label $l(v)$, which in contrast to the SPP–algorithms, gives the capacity of the (augmenting) path in $G(x)$ connecting v and s .

If node t can be labeled, a flow–augmenting path has been found and if node t cannot be labeled, then the set of labeled nodes, i.e. the set of nodes reachable from s , defines a min–cut. The following labeling method is due to Ford and Fulkerson [1962] and it works on G rather than $G(x)$.

<u>Procedure 6.5.</u>   **Labeling method for finding flow–augmenting paths**

<u>Step 0:</u> Set $p(s) := 0$, $l(s) := \infty$ and $W := \{s\}$

<u>Step 1:</u> Determine node $i \in W$, i not yet scanned.

If no such node exists, Stop: x is a max–flow and

W induces a min–cut.

If $t \in W$, Stop: flow–augmenting path P with

$\epsilon(P) = l(t)$ has been detected.

Otherwise: goto <u>Step 2</u>.

<u>Step 2:</u> Scan node i in the following way

for $(i,j) \in \delta^-(W)$ with $u_{ij} > x_{ij}$

set $l(j) := min \{l(i), u_{ij} - x_{ij}\}$

$p(j) := i$ and $W := W \cup \{j\}$

for $(j,i) \in \delta^+(W)$ with $x_{ji} > 0$

set $l(i) := min \{l(i), x_{ji}\}$

$p(j) := i$ and $W := W \cup \{j\}$

Goto <u>Step 1</u>.

The complexity of this labeling technique can be estimated as follows: During one iteration each edge has to be examined at most 2 times and each inspection may be followed by a node labeling. Thus at most $O(|E|)$ operations per iteration are necessary. Since in the case of integral capacities at most z^* augmentations are necessary the algorithm is $O(z^* \cdot |E|)$ in complexity, i.e. not polynomial in the size of the input.

Edmonds and Karp [1972] have developed refinements of the general Ford–Fulkerson–labeling method which make the procedure run in time polynomial in the size of the input. One is stated in the following

Theorem 6.8. *(Edmonds and Karp [1972])*
If in the Ford–Fulkerson–labeling method for finding a maximum flow in a network on n nodes, each flow augmentation is done along a flow–augmenting path having fewest edges, then a maximum flow will be obtained after no more than $\frac{1}{4}(n^3 - n)$ augmentations.

Flow–augmenting paths having fewest edges ("shortest augmenting paths") can easily be computed if the nodes are scanned in the same order in which they receive labels ("first labeled — first scanned" principle).

Zadeh [1972] has given a class of networks for which $O(n^3)$ augmentations are necessary, if each flow augmentation is made along a shortest augmenting path. Thus the bound given by Edmonds and Karp [1972] cannot be improved (exept for a linear scale factor).

During the last years "faster" algorithms were developed by exploiting the idea of augmenting along many paths simultaneously (cf. Dinic [1970]) and by developing more efficient data–structures, too. The running times of some recently developed max–flow algorithms are given in the following table:

Author(s):	Complexity:
Ford and Fulkerson [1956]	not necessarily polynomial
Edmonds and Karp [1969]	$O(|V| \cdot |E|^2)$
Dinic [1970]	$O(|V|^2 \cdot |E|)$
Karsanov [1974]	$O(|V|^3)$
Malhotra et al. [1978]	$O(|V|^3)$
Cherkasky [1977]	$O(|V|^2 \cdot |E|^{\frac{1}{2}})$
Galil [1978]	$O(|V|^{\frac{5}{3}} \cdot |E|^{\frac{2}{3}})$
Galil et al. [1979]	$O(|V||E|(log|V|)^2)$
Sleator [1980]	$O(|V||E|log|V|)$

Good surveys on the state–of–the art on theoretical and / or practical efficiency are Galil [1981], Cheung [1980] and Glover et al. [1979].

Chapter 7. Approaches to Min–Cost Flow Problems

In the following we analyse the network flow problem from different points of view, thereby we develop the most common approaches resp. algorithms for solving min–cost–flow and circulation problems.

An informative textbook on network flow, presenting a variety of algorithms is Jensen and Barnes [1980].

7.1. Combinatorial Analysis of Network Flow Problems — The Primal Approach

A first nontrivial question is to decide whether a given MCFP resp. CP has a feasible solution i.e. feasible flows resp. circulations exist. For the circulation problem this question is completely answered by Hoffman's existence theorem:

Theorem 7.1. *(Hoffman [1960])*
Given a network $G = (V, E)$ with lower bounds l_e and upper bounds u_e for $e \in E$, then a circulation exists iff

$$u(\delta^+(S)) \geq l(\delta^-(S)) \quad \text{for all } S \subseteq V .$$

The above theorem can be shown by reducing CP to an equivalent MCFP with lower bounds zero and then applying the existence theorem for feasible flows:

Theorem 7.2. *(Gale [1957])*
An MCFP on a network with lower capacities of zero has a feasible solution iff

$$b(S) + u(\delta^+(S)) \geq 0 \quad \text{for all } S \subseteq V .$$

Indeed the necessity of the above condition is easy to see and the sufficiency can be shown by transforming MCFP into an equivalent flow problem with only one (artificial) source and one (artificial) sink and then applying the max–flow–min–cut theorem.

Proofs for both theorems can for instance be found in Simmonard [1966]. A different proof based on the fact that MCFP has a feasible solution iff the associated LP has a feasible *basic* solution can be found in Cunningham [1976].

Finally we want to mention the following result as a special case of the theorem of Gale for the transhipment problem:

Corollary 7.3.

A transhipment problem has a feasible solution iff

$$\delta^+(S) \neq \emptyset \quad \text{for all} \quad S \subseteq V \quad \text{with} \quad b(S) < 0 \ .$$

Now with respect to a feasible circulation resp. flow $x : E \to \mathbb{N}_0$ we define the *incremental graph* $G(x) = (V, E(x))$ as we did for MFP:

$$\text{Thus let} \quad E(x) := E^+(x) \cup E^-(x) \quad \text{with}$$

$$E^+(x) := \{(i,j) \mid (i,j) \in E \ , \ x_{ij} < u_{ij}\}$$
$$E^-(x) := \{(i,j) \mid (j,i) \in E \ , \ x_{ji} > l_{ji}\} \ .$$

The edges in $E^+(x)$ are called *forward edges* and the edges in $E^-(x)$ are called *reverse edges* in $G(x)$ and a node $w \in V$ is called *reachable* from a node $v \in V$ if there exists a directed path from v to w in $G(x)$. Paths and cycles in G are called *augmenting paths* resp. *augmenting cycles* (with respect to x) if they correspond to <u>directed</u> paths resp. <u>directed</u> cycles in $G(x)$.

In $G(x)$ we define modified capacities and costs as follows:

$$c_{ij}(x) := \begin{cases} c_{ij} & \text{if } (i,j) \in E^+(x) \\ -c_{ji} & \text{if } (i,j) \in E^-(x) \end{cases}$$

$$u_{ij}(x) := \begin{cases} u_{ij} - x_{ij} & \text{if } (i,j) \in E^+(x) \\ x_{ji} & \text{if } (i,j) \in E^-(x) \end{cases}$$

$$l_{ij}(x) := 0 \quad \text{for } (i,j) \in E(x) .$$

Now let $Q = (V(Q), E(Q))$ be a directed (simple) cycle in $G(x)$ and

$$\epsilon(Q) := min \; \{u_{ij}(x) \mid (i,j) \in E(Q)\}$$

the capacity of Q .

With $Q^+ := E(Q) \cap E^+(x)$ and $Q^- := E(Q) \cap E^-(x)$ we define the incremental cost of Q by

$$c(Q) = \sum_{(i,j) \in E(Q)} c_{ij}(x) \quad = (\sum_{(i,j) \in Q^+} c_{ij} - \sum_{(i,j) \in Q^-} c_{ji}) .$$

A directed cycle Q in $G(x)$ with $c(Q) < 0$ is called a negative augmenting cycle.

For $\delta \in \{1, 2, \ldots, \epsilon(Q)\}$ let

$$y_{ij} := \begin{cases} x_{ij} + \delta & \text{for } (i,j) \in Q^+ \\ x_{ij} - \delta & \text{for } (i,j) \in Q^- \\ x_{ij} & \text{else} \end{cases} .$$

Then y is again feasible circulation with

$$\sum_{e \in E} c_e \cdot y_e = \sum_{e \in E} c_e \cdot x_e + \delta \cdot c(Q) .$$

As for the max–flow problem we will shortly write

$$y = x \oplus \delta \cdot x(Q)$$

and call this transformation an augmentation of x by Q .

Lemma 7.4.

Given two feasible circulations x and y , there exist directed cycles $Q_1, \ldots, Q_r$ in $G(x)$ and values $\delta_i \in \{1, \ldots, \epsilon(Q_i)\}$ for $i = 1, \ldots, r$ such that

$$y = x \oplus \delta_1 \cdot x(Q_1) \oplus \ldots \oplus \delta_r \cdot x(Q_r) .$$

From this lemma we can immediately deduce the following optimality criterion for circulations

Theorem 7.5.

A feasible circulation x is optimal iff there is no negative directed cycle in $G(x)$.

(This theorem is explicitly stated in Busacker and Saaty [1964] and also implicitly in Jewell [1958])

The above theorem motivates the following algorithm:

Procedure 7.1.   **Primal approach to CP**

Step 0: Determine a feasible circulation x .

Step 1: Test for existence of a negative augmenting cycle.

        If no negative augmenting cycle exists, Stop: x is optimal.

        Else given a negative augmenting cycle Q goto Step 2.

Step 2: Set $x := x \oplus \epsilon(Q) \cdot x(Q)$ and goto Step 1.

Now different methods for solving CP resp. MCFP using the primal approach have been developed. They basically vary in the way the initial circulation / flow is constructed and in the way the test for existence of negative augmenting cycles is performed.

Common approaches for Step 0 are

–   the solution of (a series of) max–flow problems

or if a basic solution has to be constructed

–   the introduction of artificial edges with large cost (Big M–method).

Klein [1967] proposed to use a matrix method for finding the shortest distance between every pair of nodes in $G(x)$ to detect negative cycles. Bennington [1973] finds negative cycles by attempting to find the shortest paths from a fixed node to all other nodes in $G(x)$. Results of computational experiments comparing this algorithm with the out–of–kilter algorithm presented, indicate that the primal approach is at least comparable to the out–of–kilter algorithm.

In the *network simplex method* (cf. Dantzig [1963], Cunningham [1976]) the incremental graph is not constructed explicitly and "candidates" for negative augmenting cycles are produced via the usual pricing–out routine of the upper-bounded simplex method. We will analyse this method in more detail in a separate section.

A special refinement of the primal approach would be to choose in each iteration a negative cycle Q with minimal (negative) cost $c(Q)$, i.e. a cycle which yields the largest marginal decrease in the objective function. This is in a certain sense the best choice for updating the solution. Yet Zadeh [1972] has given a network with 14 nodes where this approach requires an arbitrarily large number of cyclic augmentations, depending on the magnitude of the (integral) capacities only.

For the simplex–method such a worst–case network with a fixed number of nodes can not be constructed, since such a network allows only a fixed number of basic solutions. Thus, in a certain sense the simplex method is "better" than Klein's primal approach.

In fact finding the "most negative cycle" is a problem usually as hard as the travelling salesman problem. Thus such a rule cannot be implemented in practice. However the above pathological behaviour can also occur if the (negative) cycles for augmentations are chosen at random.

7.2. Linear Programming Analysis of Network Flow Problems —
The Network Simplex Method

We know that MCFP (with zero lower bounds on the edges) can be formulated as integer program

$$min \ c'x$$

$$Ax = b$$

$$0 \le x \le u$$

$$x \ \text{integer valued}$$

with A the node–edge incidence matrix of the digraph $G = (V, E)$. Since A has the "total unimodularity" property the integrality condition is superflous and thus MCFP can be solved as linear program. From LP–theory we know that an optimal (integer) solution can always be found among the set of *basic–solutions*.

The associated linear program can be brought into "standard form" by introducing slack variables $\tilde{x}$:

$$min \ c'x \quad \text{subject to}$$

$$\begin{pmatrix} A & 0 \\ E & E \end{pmatrix} \begin{pmatrix} x \\ \tilde{x} \end{pmatrix} = \begin{pmatrix} b \\ u \end{pmatrix}$$

$$x, \tilde{x} \ge 0$$

Now let B be a (column–) basis of A and let $T_B = (V, E_B)$ be the graph corresponding to B , i.e. the nodes and edges of T_B correspond to rows and columns of B respectively. The following well–known theorem is basic for any LP–approach to network flow problems:

Theorem 7.6. *If B is a basis of A then T_B is a spanning tree of G (for a proof see Dantzig [1963]) .*

Thus hereafter we will call a basic solution of MCFP also a *tree–solution*. From the above theorem we see that every basic solution of MCFP is the unique solution x of a system

$$Ax = b$$

$$x_e = u_e \quad \text{for} \ \ e \in U$$

$$x_e = 0 \quad \text{for} \ \ e \notin E(T) \cup U$$

where $T = (V, E(T))$ is a spanning tree of G and $U \subseteq E \backslash E(T)$. Hence we call (T, U) a *basis* of MCFP and we say that (T, U) is a *feasible basis* if its associated x is feasible.

An important property of a spanning tree of G is the fact that any two nodes $v, w \in V$ are connected by a <u>unique</u> path $P_{v,w}$ in the tree. Such a path can be traversed in two directions. To be able to distinguish these directions we write $P_{w,v}$ if we mean the path (direction) from w to v . An edge e contained in $P_{v,w}$ is called a *forward* (reverse) edge with respect to $P_{v,w}$ if traversing the path from v to w then $t(e)$ is reached before (after) $h(e)$.

Let us define by $P_{v,w}^+$ the set of forward edges in $P_{v,w}$ and by $P_{v,w}^-$ the set of reverse edges of $P_{v,w}$.

Throughout this section $r \in V$ will be an arbitrarily chosen fixed node in G which will serve as root for the spanning trees asociated with basic solutions.

With respect to a spanning tree $T = (V, E(T))$ and $e \in E(T)$ let

$$R(T, e) = \{v \in V \mid P_{r,v} \ \text{ does not contain } \ e\} .$$

$R(T, e)$ can also be described in the following way: Removing edge e from $E(T)$ creates two components (trees). Then $R(T, e)$ is the set of nodes which is in the same component as the root r .

84

An edge $e \in E(T)$ is said to be *directed towards* r *(away from* r*)* in T if $P_{t(e),r}$ does (not) contain e.

The following lemma gives a formula for calculating the solution x associated with a spanning tree T.

Lemma 7.7.

If x is a (feasible) tree solution associated with a spanning tree T then for any $e \in E(T)$:

$$x_e = b(R(T,e)) + u(\delta^-(R(T,e)) - u(\delta^+(R(T,e)))$$

if e is directed towards r in T resp.

$$x_e = -b(R(T,e)) - u(\delta^-(R(T,e))) + u(\delta^+(R(T,e)))$$

if e is directed away from r in T .

Proof. Follows immediately from the fact that for any feasible solution of $Ax = b$ and for all $S \subseteq V$

$$b(S) = x(\delta^+(S)) - x(\delta^-(S)) .$$

□

Given a feasible basis (T, U) and $v, w \in V$ the cost $c(P_{v,w})$ of the path directed from v to w is defined as follows

$$c(P_{v,w}) = \sum_{e \in P_{v,w}^+} c_e - \sum_{e \in P_{v,w}^-} c_e .$$

Where e is not an edge contained in $E(T)$, we define $C(T, e)$ to be the subset of E consisting of e together with the edges of the path connecting $h(e)$ and $t(e)$ in T . Again the set $C(T, e)$ may be traversed as cycle in two ways. We say that $C(T, e)$ is traversed in the direction of e (direction opposite to e) if e is a forward (reverse) edge with respect to the direction of traversal.

The cost of the cycle $C(T, e)$ is defined by

$$c(C(T, e)) = c_e + c(P_{h(e), t(e)}) \quad .$$

This cost definition can be interpreted in the following way. Let $\hat{x}$ be constructed as follows

$$\hat{x}_f = \begin{cases} x_f + \Theta & \text{for } f = e \text{ or } f \text{ forward edge in } P_{h(e), t(e)} \\ x_f - \Theta & \text{for } f \text{ reverse edge in } P_{h(e), t(e)} \\ x_f & \text{otherwise} \end{cases}$$

then $\hat{x}$ fulfills the node conservation equalities, i.e. $\hat{x}$ is a flow in G and

$$c(\hat{x}) = c(x) + \Theta \cdot c(C(T, e)) \quad .$$

Now $e \notin E(T)$ implies $x_e = 0$ or $x_e = u_e$, thus we have to distinguish two cases:

<u>Case 1:</u> If $x_e = 0$ then $C(T, e)$ (traversed in the direction of e) is an augmenting cycle in $G(x)$ iff

$$x_f < u_e \quad \text{for all forward edges in} \quad P_{h(e), t(e)}$$

$$x_f > 0 \quad \text{for all reverse edges in} \quad P_{h(e), t(e)} \quad .$$

and $c(C(T, e))$ is the incremental cost of this augmenting cycle.

<u>Case 2:</u> If $x_e = u_e$ then $C(T, e)$ (traversed in the direction opposite to e) is an augmenting cycle in $G(x)$ iff

$$x_f > 0 \quad \text{for all forward edges in} \quad P_{h(e), t(e)}$$

$$x_f < u_e \quad \text{for all reverse edges in} \quad P_{h(e), t(e)}$$

and $-c(C(T, e))$ is the incremental cost of this augmenting cycle.

If we associate a dual variable y_v with each of the nodes resp. node conservation equations of a MCFP and a dual variable w_e with the constraint $x_e \leq u_e$, the dual of MCFP is given by

$$max \sum_{v \in V} b_v y_v - \sum_{e \in E} u_e \cdot w_e \quad \text{subject to} \qquad (DMCFP)$$

$$y_{h(e)} - y_{t(e)} - w_e \leq c_e \quad \text{for } e \in E$$

$$w_e \geq 0 \quad \text{for } e \in E$$

$$y_v \quad \text{unrestricted in sign for } v \in V \quad .$$

The complementary slackness conditions of linear programming tell us that x^0 and (y^0, w^0) are optimal for MCFP and DMCFP respectively, iff

$$x_e^0 > 0 \Rightarrow y_{h(e)}^0 - y_{t(e)}^0 - w_e = c_e$$
$$w_e^0 > 0 \Rightarrow x_e^0 = u_e$$

These conditions can be reformulated as follows

$$x_e^0 = 0 \Rightarrow y_{t(e)}^0 - y_{h(e)}^0 + c_e \geq 0$$
$$x_e^0 = u_e \Rightarrow y_{t(e)}^0 - y_{h(e)}^0 + c_e \leq 0 \; .$$

Let us define

$$\bar{c}_e = y_{t(e)} - y_{h(e)} + c_e$$

the *reduced cost* of edge e . With this notation the complementary slackness conditions read

$$\bar{c}_e > 0 \Rightarrow x_e = 0$$
$$\bar{c}_e < 0 \Rightarrow x_e = u_e \; .$$

Let x be a basic feasible solution to MCFP corresponding to basis B and let $T = (V, E(T))$ be the associated spanning tree. To calculate the dual variables corresponding to B , we must set to equality the inequality dual constraints associated with basic variables i.e. edges from the tree T .
Since this is a system of linear equations with $|V|$ variables and $|V|-1$ equations, one of the variables can be fixed at an arbitrary value.

In our case, we have chosen $r \in V$ an arbitrary but fixed vertex of G to be the root of T and we set $y_r := 0$. Then the system will have a unique solution which can be evalutated recursively starting from r in the following way:

- for $e \in E(T)$ with $y_{h(e)}$ already evaluated we get

$$y_{t(e)} = y_{h(e)} - c_e$$

- for $e \in E(T)$ with $y_{t(e)}$ already evaluated we get

$$y_{h(e)} = y_{t(e)} + c_e \quad .$$

A closer look at this procedure shows that y_v , for $v \in V$, is the cost of the unique path from r to v in T , i.e. $y_v = c(P_{r,v})$.

Now consider an edge $e \notin E(T)$ and let w be the first common node of the two paths $P_{h(e),r}$ and $P_{t(e),r}$ then we get the relations

$$y_{t(e)} = y_w + c(P_{w,t(e)})$$
$$y_{h(e)} = y_w + c(P_{w,h(e)}) .$$

Moreover
$$\begin{aligned}
\bar{c}_e &= y_{t(e)} - y_{h(e)} + c_e \\
&= y_w + c(P_{w,t(e)}) - [y_w + c(P_{w,h(e)})] + c_e \\
&= c(P_{w,t(e)}) + c(P_{h(e),w}) + c_e \\
&= c(C(T,e)) .
\end{aligned}$$

From the calculation of y it is clear that $\bar{c}_e = 0$ for $e \in E(T)$ and thus the optimality conditions can equivalently be formulated in the following way:

Lemma 7.8.

Let x be a feasible basic solution with T the associated spanning tree, then x is optimal iff for any $e \notin E(T)$ the following conditions hold:

$$x_e = 0 \Rightarrow c(C(T,e)) \geq 0$$
$$x_e = u_e \Rightarrow c(C(T,e)) \leq 0 .$$

This condition is closely related to the optimality condition from section 7.1. If $x_e = 0$ then $C(T,e)$ traversed in the direction of e should not be a candidate for a negative augmenting cycle in $G(x)$ and if $x_e = u_e$ then $C(T,e)$ traversed in the direction opposite to e should not be a candidate for a negative augmenting cycle in $G(x)$.

This interpretation and formulation of the complementary slackness conditions motivates a special combinatorial or graphical formulation of the upper bounded simplex method for MCFP which we will introduce now.

<u>Procedure 7.2.</u>　　**The network simplex method**

<u>Step 0:</u>　Determine a feasible basic solution x with associated tree

$T = (V, E(T))$ and dual variables y .

<u>Step 1:</u>　Let $E^+ := \{e \notin E(T) \mid x_e = 0 \text{ and } \bar{c}_e < 0\}$

$E^- := \{e \notin E(T) \mid x_e = u_e \text{ and } \bar{c}_e > 0\}$.

If $E^+ \cup E^- = \emptyset$, Stop: x is optimal.

Otherwise choose $e \in E^+ \cup E^-$,

if $e \in E^+$, let direction of $C(T, e)$ be the same as the direction

of e

if $e \in E^-$, let direction of $C(T, e)$ be opposite to direction of e .

<u>Step 2:</u>　Let $\Theta^- := min\{x_f \mid f \text{ is a reverse edge of } C(T, e)\}$

$\Theta^+ := min\{u_f - x_f \mid f \text{ is a forward edge of } C(T, e)\}$

and $\Theta := min\{\Theta^+, \Theta^-\}$.

Let $F := \{f \mid f \text{ is a reverse edge of } C(T, e) \text{ and } x_f = \Theta\} \cup$

$\{f \mid f \text{ is a forward edge of } C(T, e) \text{ and } u_e - x_f = \Theta\}$

and choose $f \in F$

<u>Step 3:</u>　Define

$$
\hat{x}_j := \begin{cases} x_j + \Theta & j \text{ forward edge of } C(T, e) \\ x_j - \Theta & j \text{ reverse edge of } C(T, e) \\ x_j & \text{otherwise} \end{cases}
$$

$\hat{T} := (T \cup \{e\}) \backslash \{f\}$.

Calculate new dual variables $\hat{y}$ associated with $\hat{T}$.

Replace T, x and y by $\hat{T}, \hat{x}$ and $\hat{y}$ and goto <u>Step 1</u>.

Remark 1 From the definition of $\hat{x}$ it follows that

$$c'\hat{x} = c'x - \Theta \cdot |\bar{c}_e|$$

thus the objective function value strictly decreases if $\Theta \neq 0$. We will call the operation in Step 3 a *(network–simplex–)pivot* and such a pivot is said to be *degenerate* if $\Theta = 0$.

Remark 2 For the dual–update to be the performed in Step 3 the following handy formula can be used

$$\hat{y}_v = \begin{cases} y_v & \text{if } v \in R(T, f) \\ y_v + \bar{c}_e & \text{if } t(e) \in R(T, f) \text{ and } v \notin R(T, f) \\ y_v - \bar{c}_e & \text{if } h(e) \in R(T, f) \text{ and } v \notin R(T, f) \end{cases}$$

i.e. the dual variables for all the elements in $V \backslash R(T, f)$ increase or decrease by $|\bar{c}_e|$.

A characteristic of simplex–type algorithms is the flexibility of choosing the incoming non–basic variable and the leaving basic variable within certain subsets of variables. Also in the network simplex method the incoming edge e and outgoing edge f are not unique in general. Then the algorithm may be refined by prescribing a specific choice rule for the choice of e in Step 1 ("entering–edge rule") and for the choice of f in Step 3 ("leaving edge rule").

Following Bland [1977] such a refinement is referred to as a *(special) network simplex method*, whereas the *network simplex method* is the class of all such algorithms. The effect of different choice–rules with respect to the theoretical efficiency has first been evaluated by Cunningham [1976], [1979] and thereafter by Roohy–Laleh [1980]. The impact of different choice rules with respect to practical efficiency (i.e. average running time) was studied extensively by Glover, Karney and Klingman [1974], Srinivasan and Thompson [1973] , Bradley, Brown and Graves [1977] and several other authors. In the following we discuss several (recent) results concerning the theoretical efficiency of special network simplex methods.

Since the number of bases (trees) is finite the (network) simplex method is finite unless it encounters an infinite sequence of degenerate pivots caused by repeating the same sequence of bases (trees). This phenomenon is called *cycling*. In the absence of cycling the algorithm could still admit a sequence of degenerate pivots of exponential length in $|V|$, a phenomenon called *stalling*.

The first example of cycling for network flow problems was published by Gassner [1964] who gave a 4×4 assignment problem with a cycle of length 12. Cunningham and Klincewicz [1983] give an example of cycling in the network simplex method having basis size two and cycle length ten. This example is minimal in a certain sense. Of course cycling in the network simplex method can be prevented by using the special entering and leaving variable rules for the general simplex method developed by Bland [1977] :

Let $e_1, e_2, \ldots, e_n$ be an arbitrary, fixed ordering of E , then Bland's network simplex method employs the following refinements:

Refinement 1:   In Step 1, choose $e \in E^+ \cup E^-$ with the least index.

Refinement 2:   In Step 3, choose $f \in F$ with the least index.

Thus Bland's pivoting rule makes the (network) simplex method completely inflexible. In fact the rules are motivated by geometric properties of the simplex method and they do not encounter any information about the concrete example or data.

Cunningham [1976] has introduced a special subclass of "nice" MCFP-bases (trees) which he called *strongly feasible bases (trees)*. Here a feasible tree is called strongly feasible if the associated basic solution x satisfies the following property:

Every edge $f \in E(T)$ with $x_f = 0$ is directed away from r in T and every edge $f \in E(T)$ with $x_f = u_f$ is directed towards r in T .

Now the network simplex method is refined in the following way:

Refinement 1:   In Step 0, determine a strongly feasible basis x

Refinement 2: In <u>Step 2</u>, choose $f \in F$ to be the first edge in F encountered in traversing $C(T, e)$ starting at w , where $C(T, e)$ is traversed in the direction of e if $x_e = 0$ and in the direction opposite to e if $x_e = u_e$

Refinement 2 ensures that the tree associated with y is again a strongly feasible tree. The network simplex method with these refinements is called the *strongly feasible network simplex method*. The following result is crucial for the property of non–cycling.

Lemma 7.9. *(Cunningham [1976])*

Let T be a strongly feasible tree and $T' = T \backslash \{e\} \cup \{f\}$ the strongly feasible tree obtained from T by a degenerate pivot of the strongly feasible network simplex method and let y, y' be the associated dual solutions. Then

$$\sum_{v \in V} y'_v < \sum_{v \in V} y_v ,$$

thus no tree can be repeated during a sequence of degenerate pivots.
Hence the method cannot cycle.

The important advantage of Cunningham's refinement over Bland's rule is the fact that "strong feasibility" is assured by only specifying the choice for the leaving edge. The user still has full freedom for choosing the incoming edge among the set $E^+ \cup E^-$. Thus the strongly feasible network simplex method can be implemented with any of the empirically efficient choice rules for the incoming edge as Dantzig's " most negative rule" etc. (cf. Mulvey [1978], Bradley et al. [1977]).

If a strongly feasible tree to initiate the algorithm is not available, one can be constructed using "artificial" edges having sufficiently large cost (Big M–method). Cunningham [1976] , [1979] has given two procedures to convert any feasible tree into a strongly feasible tree.

Cunningham [1979] has shown that without specific choices for the incoming edge the strongly feasible network simplex method as well as the network simplex method with Bland's rule may stall. Now it is an immediate question whether special stalling–preventing choice rules for the incoming edge in the strongly feasible network simplex method can be given to prevent stalling.

Avis and Chvatal [1978] have shown that if cycling and stalling could be prevented for the general linear programming simplex method, then it becomes a "good" algorithm in the sense of Edmonds, i.e. it becomes a polynomial algorithm with respect to the problem size (number of constraints and number of variables).

In their proof, Avis and Chvatal transform the given LP problem into an equivalent problem with right–hand–side vector zero. Now their transformation would not preserve the network structure when applied to MCFP. Thus from their results we cannot conclude that it is sufficient to prevent stalling (and cycling) to make the network simplex method a "good" algorithm.

Several stalling preventing rules have been developed by Cunningham [1979] and Roohy–Loleh [1980]. A major breakthrough was the discovery of a choice rule for the entering edge which in the case of an assignment problem gives a polynomial algorithm. To our knowledge this has been the first and so far only example for a polynomial simplex method.

7.3. Combinatorial Analysis of DCP — A Unifying Dual Approach

In this section we start from the most general network flow problem — the circulation problem with lower bounds, i.e. the linear program

$$min \sum_{e \in E} c_e \cdot x_e \quad \text{subject to} \quad \text{(CP)}$$

$$x(\delta^+(v)) - x(\delta^-(v)) = 0 \quad \text{for } v \in V$$

$$l_e \leq x_e \leq u_e \quad \text{for } e \in E$$

If we associate a dual variable y_v with each of the nodes resp. node conservation equations, a dual variable λ_e with the constraint $x_e \geq l_e$ a dual variable μ_e with the constraint $x_e \leq u_e$ which we transform into $-x_e \geq -u_e$, then the dual of the circulation problem is given by

$$\text{maximize} \sum_{e \in E} l_e \cdot \lambda_e - \sum_{e \in E} u_e \cdot \mu_e \quad \text{subject to} \quad \text{(DCP)}$$

$$- y_{t(e)} + y_{h(e)} + \lambda_e - \mu_e = c_e \quad \text{for } e \in E$$

$$\lambda_e, \mu_e \geq 0 \quad \text{for } e \in E$$

$$y_v \quad \text{unrestricted in sign} \quad \text{for } v \in V$$

Now it is easy to see that DCP has always a feasible solution. Suppose we select any set of y_v 's then the dual constraint for an edge becomes

$$\lambda_e - \mu_e = c_e + y_{t(e)} - y_{h(e)}$$

which can be satisfied by

$$\lambda_e := max\{0, \ c_e + y_{t(e)} - y_{h(e)}\}$$

$$\mu_e := - min\{0, \ c_e + y_{t(e)} - y_{h(e)}\}$$

Again we define

$$\bar{c}_e := c_e + y_{t(e)} - y_{h(e)}$$

the *reduced cost* of edge e and

$$\bar{c}_e^+ := max\{0, \, \bar{c}_e\}$$
$$\bar{c}_e^- := min\{0, \bar{c}_e\} \, .$$

Then DCP becomes

$$max \sum_{e \in E} l_e \cdot \bar{c}_e^+ + \sum_{e \in E} u_e \cdot \bar{c}_e^- \quad \text{subject to}$$

$$- y_{t(e)} + y_{h(e)} + \bar{c}_e = c_e \quad \text{for } e \in E$$

$$y_v \quad \text{unrestricted in sign} \quad \text{for } v \in V$$

$$\bar{c}_e \quad \text{unrestricted in sign} \quad \text{for } e \in E$$

The complementary slackness conditions for optimality have the form

$$(x_e - l_e) \cdot \bar{c}_e^+ = 0 \quad \text{for } e \in E$$
$$(u_e - x_e) \cdot \bar{c}_e^- = 0 \quad \text{for } e \in E$$

or equivalently

$$\bar{c}_e > 0 \quad \Rightarrow \quad x_e = l_e$$
$$\bar{c}_e = 0 \quad \Rightarrow \quad l_e \le x_e \le u_e \quad \text{for } e \in E$$
$$\bar{c}_e < 0 \quad \Rightarrow \quad x_e = u_e$$

Thus if the reduced cost of an edge is negative (positive) the flow on this edge has to be at its upper (lower) level.

With respect to a set of dual variables and its associated reduced costs we define modified bounds

$$\bar{l}_e = \begin{cases} l_e & \text{if } \bar{c}_e \ge 0 \\ u_e & \text{if } \bar{c}_e < 0 \end{cases}$$

$$\bar{u}_e = \begin{cases} l_e & \text{if } \bar{c}_e > 0 \\ u_e & \text{if } \bar{c}_e \le 0 \end{cases}$$

Any circulation in G with respect to these modified lower and upper bounds is an optimal circulation for the original problem.

According to Hoffman's existence theorem such a circulation exists iff

$$\bar{u}(\delta^-(S)) \geq \bar{l}(\delta^+(S)) \quad \text{for all } S \subseteq V .$$

For any set $S \subset V$ we define

$$\gamma(S) = \bar{l}(\delta^+(S)) - \bar{u}(\delta^-(S))$$

and we call a set $S \subseteq V$ with $\gamma(S) > 0$ an *improving set*.

For improving sets $S \subseteq V$ we define

$$\epsilon(S) := min \begin{cases} -\infty \\ \{-\bar{c}_e \mid e \in \delta^+(S) , \ \bar{c}_e < 0\} \\ \{\bar{c}_e \mid e \in \delta^-(S) , \ \bar{c}_e > 0\} \end{cases} .$$

Now setting

$$\hat{y}_v := \begin{cases} y_v + \epsilon(S) & \text{for } v \in S \\ y_v & \text{otherwise} \end{cases}$$

gives a new dual solution with a larger (dual) objective function value as can be seen easily. In fact, the dual objective function value invreases by $\gamma(S) \cdot \epsilon(S)$.

The above observations motivate the following approach:

Procedure 7.3.   **Dual approach to CP**

Step 0: Determine dual variables y_v for $v \in V$ and calculate the associated reduced costs $\bar{c}_e$ for $e \in E$.

Step 1: Test for existence of an improving set $S \subseteq V$.
         If no such set exists, construct the associated optimal circulation and stop.
         Else given an improving set S , goto Step 2

Step 2: Set $y_v = y_v + \epsilon(S)$ for $v \in S$ and goto Step 1.

Hassin [1983] has shown that some of the most common existing network flow algorithms

- the out–of–kilter algorithm

- the dual simplex method

- the primal–dual algorithm

fit into the above scheme. They only vary in the way they search for improving sets.

In fact Zadeh [1980] has shown that the above mentioned algorithms applied to MCFP are equivalent in the sense that they produce (modulo ties) the same sequence of intermediate dual solutions when they are started from the same initial dual solution, $y_v = 0$ for $v \in V$ for instance. Using Hassin's interpretation this is due to the fact that (modulo ties) the same sequence of improving sets is determined.

In the following we outline two well–known representatives out of the class of "dual" network flow algorithms.

The out–of–kilter–method

The out–of–kilter–method was independently introduced by Fulkerson [1961], Minty [1960] and Yakovleva [1959]. Efficient implementations are described by Barr et al. [1974] and Aashtiani and Magnanti [1976].

The procedure can be started with any circulation x and any set of dual variables y_v , $v \in V$. Then an edge violating any of the optimality conditions

(I1) $l_e \leq x_e \leq u_e$ for $e \in E$

(I2) $\bar{c}_e > 0 \;\Rightarrow\; x_e = l_e$

(I3) $\bar{c}_e < 0 \;\Rightarrow\; x_e = u_e$

is said to be an *out–of–kilter edge* and any edge fulfilling the above conditions is said to be an *in–kilter edge*.

For out–of–kilter edges $e \in E$ we can distinguish six possible states

97

(O1) $\bar{c}_e > 0$ and $x_e < l_e$

(O2) $\bar{c}_e > 0$ and $x_e > l_e$

(O3) $\bar{c}_e = 0$ and $x_e < l_e$

(O4) $\bar{c}_e = 0$ and $x_e > u_e$

(O5) $\bar{c}_e < 0$ and $x_e < u_e$

(O6) $\bar{c}_e < 0$ and $x_e > u_e$

Now the *kilter number* $K(e)$ for an edge $e \in E$ is defined as follows:

$$K(e) = \begin{cases} l_e - x_e & \text{if } \bar{c}_e \geq 0 \text{ and } x_e < l_e \\ \bar{c}_e(x_e - l_e) & \text{if } c_e > 0 \text{ and } x_e > l_e \\ x_e - u_e & \text{if } c_e \leq 0 \text{ and } x_e > u_e \\ \bar{c}_e(u_e - x_e) & \text{if } c_e < 0 \text{ and } x_e < u_e \\ 0 & \text{else} \end{cases}$$

A node j is said to be *reachable* from node i if either

(R1) $\bar{c}_{ij} \leq 0$ for edge $(i,j) \in E$ and $x_{ij} < u_{ij}$

(R2) $\bar{c}_{ij} > 0$ for edge $(i,j) \in E$ and $x_{ij} < l_{ij}$

(R3) $\bar{c}_{ji} \geq 0$ for edge $(j,i) \in E$ and $x_{ji} > l_{ji}$

(R4) $\bar{c}_{ji} < 0$ for edge $(j,i) \in E$ and $x_{ji} > u_{ji}$

This relation induces a directed graph $G(y) = (V, E(y))$ where $(i,j) \in E(y)$, if j is reachable from i. Every edge $(i,j) \in E(y)$ corresponds either to an edge $(i,j) \in E$ in which case it is called a forward edge in $E(y)$ or an edge $(j,i) \in E$ in which case it is called a reverse edge in $E(y)$. Note that every out–of–kilter edge induces an edge in $G(y)$.

Now let edge $(k,l) \in E$ be an out–of–kilter edge fulfilling (R1) or (R2) or edge $(l,k) \in E$ be an out–of–kilter edge fulfilling (R3) or (R4). In either case l is reachable from k and thus $(k,l) \in E(y)$.

Assume that exists a directed path $P_{l,k}$ in $G(y)$ connecting l to k. Then this path together with edge (k,l) forms a directed cycle C in $G(y)$.

Now the "flows" x_{ij} on forward edges (i,j) in C become $x_{ij} + \Theta$ and the "flows" x_{ji} on reverse edges (i,j) in C become $x_{ji} - \Theta$ where

$$\Theta := min\{\Theta_{ij} \mid (i,j) \in E(y) \text{ edges of } C\} \text{ and}$$

$$\Theta_{ij} := \begin{cases} u_{ij} - x_{ij} & \text{if } (i,j) \text{ forward edge of } C \text{ and } \bar{c}_{ij} \leq 0 \\ l_{ij} - x_{ij} & \text{if } (i,j) \text{ forward edge of } C \text{ and } \bar{c}_{ij} > 0 \\ x_{ji} - l_{ji} & \text{if } (i,j) \text{ reverse edge of } C \text{ and } \bar{c}_{ji} \geq 0 \\ x_{ji} - u_{ji} & \text{if } (i,j) \text{ reverse edge of } C \text{ and } \bar{c}_{ji} < 0 \end{cases}$$

By this "augmentation" a new circulation is constructed where every in–kilter–edge with respect to the old circulation stays in kilter, no kilter number is increased and the kilter number for at least one out–of–kilter edge of the cycle C is reduced by Θ .

If no such directed path from l to k exists in $G(y)$, let S be the set of nodes which are reachable from l in $G(y)$. Then S is an improving set and we would perform a transformation.

To show that S is an improving set, we have to distinguish two classes of edges.

For $(i,j) \in E$ with $i \in S$ and $j \notin S$ either

$$\bar{c}_{ij} \leq 0 \quad \text{and} \quad x_{ij} \geq u_{ij} \quad \text{or}$$

$$\bar{c}_{ij} > 0 \quad \text{and} \quad x_{ij} \geq l_{ij} \quad \text{holds,}$$

and for $(j,i) \in E$ with $i \in S$ and $j \notin S$ either

$$\bar{c}_{ji} \geq 0 \quad \text{and} \quad x_{ji} \leq l_{ji} \quad \text{or}$$

$$\bar{c}_{ji} < 0 \quad \text{and} \quad x_{ji} \leq u_{ji} \quad \text{holds.}$$

Hence $x(\delta^-(S)) \geq \bar{u}(\delta^-(S))$ and $x(\delta^+(S)) \leq \bar{l}(\delta^+(S))$.

Since x is a circulation we have $x(\delta^-(W)) = x(\delta^+(W)$ for all $W \subseteq V$ and thus

$$\bar{u}(\delta^-(S)) \leq \bar{l}(\delta^+(S)) \qquad \text{holds .}$$

However at least the edge (k,l) resp. edge (l,k) has positive kilter number and therefore strict inequality holds resp. $\gamma(S) > 0$.

Note that in an implementation of the out–of–kilter method, a path P connecting l and k in $G(y)$ resp. an improving set S can be determined by Ford and Fulkerson's labeling method for finding augmenting paths in the max–flow approach (Procedure 6.5.). Thereby the capacity Θ is computed simultaneously. The finiteness of the out–of–kilter method can be shown as follows:

Whenever a directed path in $G(y)$ connecting l and k is constructed, the kilter number of edge (k,l) resp. edge (l,k) is reduced by the positive amount $\epsilon(P)$, which is an integer if the initial circulation was integer valued. Hence after a finite number of "augmentations" the edge $(k,l) \in E$ will become an in–kilter–edge. A "dual change" does not effect the reduced cost of edges in $\gamma(S)$ and $\gamma(V \setminus S)$ and for at least one edge which was defining $\epsilon(S)$ the reduced cost becomes zero. Thus either the number of nodes which are reachable from l increases after the dual change, or if this set S does not change, then the number of edges defining $\epsilon(S)$ decreases. Hence only a finite number of dual changes is possible without either finding a directed path connecting l and k or obtaining $\epsilon(S) = \infty$, in which case no feasible circulation exists.

The primal–dual algorithm

This approach was originally developed by Ford and Fulkerson [1957] for the capacitated Hitchcock problem and can be viewed as a generalization of Kuhn's famous Hungarian method for solving assignment problems (cf. Kuhn [1955] , [1956]). The method is started with any dual solution y and the edge set E is partitioned into three sets relative to y

$$E^-(y) := \{e \in E \mid \bar{c}_e < 0\}$$
$$E^0(y) := \{e \in E \mid \bar{c}_e = 0\}$$
$$E^+(y) := \{e \in E \mid \bar{c}_e > 0\} .$$

Then the flow values are set as follows

$$x_e = u_e \quad \text{for} \quad e \in E^-(y)$$
$$x_e = l_e \quad \text{for} \quad e \in E^+(y) .$$

If now the flow variables x_e for $e \in E^0(y)$ can be selected within the intervals $[l_e, u_e]$ such that all flow conservation equalities are fulfilled, then an optimal circulation has been constructed.

For this purpose we define for $v \in V$:

$$b_v(y) = -\left(\sum_{e \in \delta^+(v) \cap E^-(y)} u_e + \sum_{e \in \delta^+(v) \cap E^+(y)} l_e - \sum_{e \in \delta^-(v) \cap E^-(y)} u_e - \sum_{e \in \delta^-(v) \cap E^+(y)} l_e \right)$$

Then the selection of the flow values x_e for $e \in E^0(y)$ is done by solving the following restricted primal flow problem

$$\min \sum_{v \in V} s_v^+ + \overline{s_v^-} \quad \text{subject to} \tag{RP}$$

$$x(\delta^+(v) \cap E^0(y)) - x(\delta^-(v) \cap E^0(y)) + s_v^+ - s_v^- = b_v(y) \quad \text{for} \quad v \in V$$

$$l_e \leq x_e \leq u_e \quad \text{for} \quad e \in E^0(y)$$

$$s_v^+, s_v^- \geq 0 \quad \text{for} \quad v \in V .$$

The above problem can again be transformed into an equivalent problem with lower bounds zero:

$$\min \sum_{v \in V} s_v^+ + s_v^- \quad \text{subject to}$$

$$\tilde{x}(\delta^+(v) \cap E^0(y)) - \tilde{x}(\delta^-(v) \cap E^0(y)) + s_v^+ - s_v^- = \bar{b}_v(y) \quad \text{for} \quad v \in V$$

$$0 \leq \tilde{x}_e \leq u_e - l_e \quad \text{for} \quad e \in E^0(y)$$

$$s_v^+, s_v^- \geq 0 \quad \text{for} \quad v \in V$$

where $\bar{b}_v(y) = b_v(y) - l(\delta^+(v) \cap E^0(y)) + l(\delta^-(v) \cap E^0(y))$.

The latter problem can be formulated as a max–flow problem by adding a dummy source s and a dummy sink t and edges

$$(s, v) \quad \text{with} \quad u_{sv} = -\bar{b}_v(y) \quad \text{if} \quad \bar{b}_v(y) < 0$$

$$(v, t) \quad \text{with} \quad u_{vt} = \bar{b}_v(y) \quad \text{if} \quad \bar{b}_v(y) > 0$$

and then treating all nodes $v \in V$ as transhipment nodes. This max–flow problem can now be solved by the labeling method of Ford and Fulkerson (cf. section 6.2.).

Now given a max–flow $\tilde{x}$ in this extended network we define

$$x_e = \tilde{x}_e + l_e \quad \text{for } e \in E^0(y) \quad \text{and}$$

$$s_v^+ = max \ \{0, b_v(y) - x(\delta^+(v) \cap E^0(y)) + x(\delta^-(v) \cap E^0(y))\}$$

$$s_v^- = - min \ \{0, b_v(y) - x(\delta^+(v) \cap E^0(y)) + x(\delta^-(v) \cap E^0(y))\}$$

This gives an optimal solution for RP. Moreover if the maximal flow $\tilde{x}$ has value $\tilde{z} = \sum_{\tilde{b}_v > 0} \tilde{b}_v(y)$, then

$$\sum_{v \in V} s_v^+ + s_v^- = 0$$

and the vector $(x_e \mid e \in E)$ is an optimal solution for the original circulation problem. Otherwise the set S of labeled nodes is an improving set and we would use this set to update the dual solution.

Now let us assume that our original problem to be solved is a MCFP with zero lower bounds and the initial dual solution is $y \equiv 0$.
In this special case we have

$$\bar{c}_e \geq 0 \quad \text{for} \quad e \in E$$

and after extending the network by a dummy source and a dummy sink and the appropriate set of artificial edges the restricted primal problem is essentially the problem of finding a max–flow in the graph using edges with reduced cost zero only. We will further analyse this special case of the primal–dual approach to MCFP in section 8.1.

Zadeh [1979] has proposed a "simple alternative to replace the out–of–kilter algorithm":

Given an initial circulation x which may violate upper and lower bounds, it suffices to find a circulation y such that $x + y$ is optimal. But this is a circulation problem in which the initial flows are zero, and therefore violate only lower bounds. Using the standard transformation this circulation problem can be formulated as a MCFP with zero lower bounds. Then the steps of the out–of–kilter–algorithm can be interpreted within this environment.

Now assume again that the original problem is already stated as MCFP with lower bounds zero and the initial dual solution is $y \equiv 0$. In this case it is easy to see from the above that the out–of–kilter–method boils down to the primal–dual approach which shows the "partial" or "near equivalence" of these two dual algorithms.

7.4. Extreme Flows — The Shortest Augmenting Path Method

In this section we treat the special case of MCFP with only one source s and only one sink t and zero lower bounds. Then we call a mapping x a s–t–flow if it fulfills

$$x(\delta^+(v)) - x(\delta^-(v)) = \begin{cases} -z & \text{for } v = s \\ z & \text{for } v = t \\ 0 & \text{for } v \in V\backslash\{s,t\} \end{cases}$$

$$0 \le x_e \le u_e \quad \text{for all} \quad e \in E$$

$$x \quad \text{integer valued} .$$

Here $z(x) := z$ is called the *flow value* of x .

Then a s–t–flow $\bar{x}$ is called *extreme* if it is of minimum cost among flows with value $z(\bar{x})$, i.e. $\bar{x}$ is an optimal solution of the program

$$min \; \{\sum_{e \in E} c_e \cdot x_e \mid x \text{ is a } s\text{–}t\text{–flow with } z(x) = z(\bar{x})\} .$$

Now we can use the results developed in the section above to characterize extreme flows via combinatorial and linear programming related criteria:

Lemma 7.10.

Let x be a s–t–flow, then the following are equivalent

(i) x is extreme,

(ii)   *every directed cycle in $G(x)$ has nonnegative weight,*

(iii)   *there exists a "labeling function" $y : V \to \mathbb{Z}$ s.t.*

$$y_i - y_j + c_{ij} > 0 \quad \Rightarrow \quad x_{ij} = 0$$
$$y_i - y_j + c_{ij} < 0 \quad \Rightarrow \quad x_{ij} = u_{ij}$$

A flow x and a labeling function y together fulfilling the properties from (iii) are called a *compatible pair*. The key result concerning extreme flows is the following

Theorem 7.11.   *If x is an extreme flow of value z and if P is a shortest augmenting path in $G(x)$ with capacity $\epsilon(P)$ then for $\delta \in \{1, \ldots, \epsilon(P)\}$*

$$\bar{x} = x \oplus \delta \cdot x(P)$$

is an extreme flow of value $z + \delta$.

This basic theorem is attributed to Jewell [1958], Busacker and Gowen [1961] and Iri [1960]. A simple proof can also be found in Papadimitriou and Steiglitz [1982].

The above theorem suggests the following method for solving MCFP :

"Starting with an extreme flow x^0 , compute a sequence of extreme flows $x^1, x^2, \ldots, x^h, x^{h+1}, \ldots$ obtaining x^{h+1} from x^h by augmenting along a shortest augmenting path from s to t in $G(x^h)$."

Due to the fact that x^h is extreme it follows that $G(x^h)$ does not contain negative cycles and hence the shortest path problems can be solved by efficient shortest path algorithms.

Independently Edmonds and Karp [1972] and Tomizawa [1972] have developed a modified approach where in each iteration a shortest path problem on a graph with nonnegative edge–weights has to be solved. For this problem the highly efficient Dijkstra–method [cf. section 6.1.] can be used.

Recall that given a flow x the incremental graph $G(x) = (V, E(x))$ was defined via $E(x) := E^+(x) \cup E^-(x)$ with

$$E^+(x) := \{(i,j) \mid (i,j) \in E \text{ with } x_{ij} < u_{ij}\}$$
$$E^-(x) := \{(i,j) \mid (j,i) \in E \text{ with } x_{ji} > 0\},$$

and the incremental cost of an edge $(i,j) \in E(x)$ was defined by

$$c_{ij}(x) = \begin{cases} c_{ij} & \text{if } (i,j) \in E^+(x) \\ -c_{ji} & \text{if } (i,j) \in E^-(x) . \end{cases}$$

Now assume that (x, y) is a compatible pair, then we assign each edge $(i,j) \in E(x)$ a "reduced incremental cost"

$$\bar{c}_{ij}(x) = y_i - y_j + c_{ij}(x)$$

Then

$$(i,j) \in E^+(x) \text{ implies } x_{ij} < u_{ij} \text{ hence}$$
$$y_i - y_j + c_{ij} = y_i - y_j + c_{ij}(x) \geq 0$$
$$(i,j) \in E^-(x) \text{ implies } x_{ji} > 0 \text{ hence}$$
$$y_j - y_i + c_{ji} = y_j - y_i - c_{ij}(x) \leq 0 .$$

Thus $\bar{c}_{ij}(x) \geq 0$ for all $e \in E$.

Moreover the following basic relations hold

(i)   If $Q = (V(Q), E(Q))$ is a directed cycle in $G(x)$, then

$$\sum_{(i,j) \in E(Q)} c_{ij}(x) = \sum_{(i,j) \in E(Q)} \bar{c}_{ij}(x) .$$

(ii)   If $P = (V(P), E(P))$ is a directed path in $G(x)$ from i^* to j^* , then

$$\sum_{(i,j) \in E(P)} \bar{c}_{ij}(x) = y_{i^*} - y_{j^*} + \sum_{(i,j) \in E(P)} c_{ij}(x) .$$

Thus $G(x)$ has a negative cycle with respect to the weights $c_{ij}(x)$ iff it has a negative cycle with respect to the weights $\bar{c}_{ij}(x)$. Also, P is a shortest s–t–path in $G(x)$ with respect to the weights $c_{ij}(x)$ if P is a shortest s–t–path in $G(x)$ with respect to the weights $\bar{c}_{ij}(x)$.

Since $\bar{c}_{ij}(x) \geq 0$ for $(i,j) \in E(x)$ a shortest flow augmenting path can be found by the highly efficient Dijkstra–labeling method when using the weights $\bar{c}_{ij}(x)$ instead of $c_{ij}(x)$. Moreover, after the determination of the shortest augmenting path and the flow change the node–labels of the Dijkstra–method can be used to construct a new labeling function which is compatible with the new flow.

The complete algorithm for finding a min–cost max–s–t–flow is as follows:

<u>Procedure 7.4.</u>   **Shortest augmenting path algorithm**

<u>Step 0</u>: Determine compatible pair (x^0, y^0) and set $h = 0$.

       $(x^0 \equiv 0, y^0 \equiv 0$ for instance)

<u>Step 1</u>: Determine the shortest s–t–path P in $G(x)$ with respect to the nonnegative weights $\bar{c}_{ij}(x^h)$ using Dijkstra's labeling technique. If several shortest path's exist, choose one with the fewest edges and set

    $x^{h+1} := x^h \oplus \epsilon(P) \cdot x(P)$ and goto <u>Step 2</u>.

If no augmenting path exists : Stop x^h is a min–cost–max–flow.

<u>Step 2</u>: If $l^h(v)$ denotes the cost of the shortest path from s to v in $G(x^h)$ with respect to weights $\bar{c}_{ij}(x^h)$ set

    $y_v^{h+1} := y_v^h + l^h(v)$ for $v \in V$.

Set $h := h + 1$ and goto <u>Step 1</u>.

Throughout the procedure (i.e. for all h) $y_s^h = 0$ and y_v^h gives the length of a shortest path from s to v in $G(x^h)$, since the existence of a path from x to v in $G(x^h)$ with shorter length would show the existence of a negative cycle in $G(x^h)$ and contradict the fact that x^h is extreme. Moreover, y^{h+1} and x^{h+1} are compatible and $y_v^{h+1} \geq y_v^h$ for all $v \in V$.

It is evident how this method has to be modified if a min–cost–flow with fixed flow value z (possibly less than then maximal possible flow value z^*) is required.

Edmonds and Karp [1972] have given two bounds on the number of flow augmentation required by the shortest augmenting path method:

Theorem 7.12. *If all capacities are integer, then the computation terminates after at most z^* flow augmentations. If all costs c_{ij} are integers less than or equal to an integer K , then the computation terminates after at most $\frac{1}{4}(n^3 - n)(n - 1) \cdot K + 1$ flow augmentations.*

Zadeh [1973] has given a "bad" class of networks with $2n + 2$ nodes on which the shortest augmenting path method requires $2^n + 2^{n-2} - 2$ iterations i.e. flow augmentations.

Although the concept of extreme flows resp. the shortest augmenting path approach is not a "good" method in the sense of Edmonds, the idea has lead to the first (and so far only) polynomial network flow algorithm (besides the "uncombinatorial" ellipsoid method). This modification, called "scaling–technique", will be introduced in section 11.2.

Chapter 8. Near Equivalence of Network Flow Algorithms

So far we have introduced the most common algorithms for solving network flow problems together with their theoretical background and motivation. There has been quite a number of articles which investigated the practical efficiency of the different approaches.

In the 50's the simplex method was the only tool for solving network flow problems, then in the 60's after the development of the out–of–kilter method and its variants those "dual" methods were considered to be the most efficient approach. In the beginning of the 70's with the work of Glover et al. [1974] and some other research groups on efficient data–structures for storing network–LP–bases as well as clever starting and pivoting strategies, the simplex method became the winner again. Today we can see that the race is over although there are still some "local activities" going on. Rather efficient — and expensive — software packages for solving network problems habe been produced and were extensively described and compared in literature.

At the same time when the discussion on the practical efficiency of different approaches started, Zadeh [1973] besides others investigated the question about the theoretical efficiency by producing bad–examples for all standard procedures (besides the scaling technique which we did not discuss so far). In two reports (Zadeh [1972], [1980]) he claimed that all the standard network algorithms

- out–of–kilter

- primal and dual simplex

- primal–dual

- complementary pivot

can be viewed as special implementations of the

- shortest augmenting path approach

if starting conditions are set appropriately. To our knowledge these reports and the results have never been published and therefore even today this observation is not at all widely known.

Indeed Zadeh's results — which several other researchers may also have (at least partially) discovered — indicate that the shortest augmenting path concept can be viewed as the backbone of all clever approaches. It also is the basis for the only (combinatorial) approach to network flow problems which is polynomial. Moreover this knowledge can also help to understand results from computational studies on special pivoting strategies etc.

This result also has another more "philosophical" impact on the theory of algorithms for discrete optimization problems.

One of the most simple — at the same time most attractive — algorithmic principles in combinatorial optimization is the *greedy principle* (cf. Edmonds [1971]. Usually this procedure is interpreted as a myopic decision rule where the solution is developed or composed through a sequence of "local optimal" solutions. Combinatorial structures are studied where such decisions on early stages never have to be changed lateron (cf. Korte and Lovasz [1982]).

Now an extreme flow can be viewed as such a local optimal "solution" — there is no cheaper way to send the same amount of flow from s to t . Then, if the demand at node t is raised by one unit, the shortest augmenting path tells us how to (partially) correct our previous decision to get to the new (local) optimum. Here the shortest augmenting path approach — and hence the other common methods too — mimic this decision process, where the demand at the sink is increased constantly.

In a later section we will show how this *myopic rule* (shortest augmenting path) combined with an *anticipant organisation* (scaling technique) will lead to a combinatorially motivated, efficient network flow algorithm.

In the following we will demonstrate the *near equivalence* of the primal simplex method, the primal–dual method and the shortest augmenting path method for the MCFP with one source and one sink.

8.1. Equivalence of the Primal–Dual Method and the Shortest Augmenting Path Method

In the following we assume that the shortest augmenting path method and the primal dual method are started from the same initial flow $x^0 \equiv 0$ and dual variables / node numbers $y^0 \equiv 0$. To show the (partial) equivalence we break both methods into comparable *phases*.

By such a phase we mean the sequence of steps performed between two augmentations. Thus the k–th phase starts from a flow x^k and node numbers y^k and produces an altered flow x^{k+1} and altered dual variables y^{k+1} . We will show now that given (x^k, y^k) both methods will construct (modulo) ties the same (x^{k+1}, y^{k+1}) and this proves the near equivalence of both approaches.

Review of the shortest augmenting path method

A phase of the shortest augmenting path method consists of the application of the Dijkstra labeling method to find the shortest s–t–path in $G(x^k)$ with respect to the costs $\bar{c}_{ij}(x^k) := c_{ij}(x^k) + y_i^k - y_j^k$, followed by a "dual update":

Procedure 8.1.   **Shortest augmenting path computation**

Step 0: Set $S := \{s\}$ and $l(s) := 0$

$$l(j) := \begin{cases} \bar{c}_{sj}(x^k) & \text{for } (s,j) \in E(x^k) \\ \infty & \text{otherwise} \end{cases}$$

Step 1: Determine $w \in V \backslash S$ with $l(w) = min \ \{l(v) \mid v \in V \backslash S\}$.

     If $l(w) = \infty$, Stop: no augmenting path exists, x^k is a max–flow.

     If $w = t$ goto Step 3,

     otherwise goto Step 2.

Step 2: Set $S \cup \{w\}$, and for all $v \in V \backslash S$ set

$$l(v) := min\{l(v), l(w) + \bar{c}_{wv}(x^k)\} .$$

Goto Step 1

Step 3: Let P be the s–t–path found in Step 1.

Set $x^{k+1} := x^k + \epsilon(P) \cdot x^k(P)$ and

$$\text{set} \quad y_j^{k+1} := \begin{cases} y_j^k + l(j) & \text{for } j \in S \\ y_j^k + l(t) & \text{for } j \in V \backslash S . \end{cases}$$

Review of the primal dual method

In section 7.3. we have already shown that for this special kind of problem and with this start condition the primal subproblems are to find max s–t–flows in G using edges with reduced cost zero only and the dual subproblem is to update the dual variables via an improving set which is determined by a min-cut.

At a general step with the s–t–flow x^k at hand let y_i^k , $i \in V$, be the dual variables at the beginning of the current phase and let Δy_i^k , $i \in V$, such that $y_i^k + \Delta y_i^k$, $i \in V$, is the present node labeling. Then an edge $(i,j) \in E(x^k)$ is *admissible* if

$$(y_i^k + \Delta y_i^k) + c_{ij}(x^k) - (y_j^k + \Delta y_j^k) = 0 .$$

Now the following labeling procedure alternatively searches for an s–t–path in $G(x^k)$ or uses the set S of "labeled" nodes as improving set to update the node labeling.

Procedure 8.2.   **Primal–dual labeling procedure**

Step 0: Set $S := \{s\}$ and $\Delta y_i^k := 0$ for all $i \in V$

Step 1: Let S be the set of all nodes reachable from s in $G(x^k)$

by a path of admissible edges, i.e. edges $(i,j) \in E(x^k)$ s.t.
$$(y_i^k + \Delta y_i^k) + c_{ij}(x^k) - (y_j^k + \Delta y_i^k) = 0$$
If $t \in S \to$ goto Step 3.

111

Otherwise determine

$$\delta := min \ \{(y_i^k +\triangle y_i^k) + c_{ij}(x^k) - (y_j^k +\triangle y_j^k) \mid i \in S, \ j \in V\backslash S\}$$

and goto <u>Step 2</u>

<u>Step 2:</u> Set $\triangle y_j^k := \triangle y_j^k + \delta$ for all $j \in V\backslash S$

and goto <u>Step 1.</u>

<u>Step 3:</u> Let P be the $s\text{--}t\text{--}$path found in <u>Step 1</u>

Set $x^{k+1} := x^k + \epsilon(P) \cdot x^k(P)$

and set $\quad y_i^{k+1} := \begin{cases} y_i^k +\triangle y_i^k & \text{if } \triangle y_i^k \leq \triangle y_t^k \\ y_i^k +\triangle y_t^k & \text{otherwise} \end{cases}$

To demonstrate the equivalence of both methods we rewrite the formula for the $\delta-$ calculation in the primal–dual method as follows

$$\delta = min \ \{(y_i^k +\triangle y_i^k) + c_{ij}(x^k) - (y_j^k +\triangle y_j^k) \mid i \in S, \ j \in V\backslash S\}$$

$$= min \ \{\triangle y_i^k + \bar{c}_{ij}(x^k) -\triangle y_j^k \mid i \in S, \ j \in V\backslash S\} \ .$$

Now $\triangle y_j^k$ is constant for all $j \in V\backslash S$ and thus the algorithm determines a node $j \in V\backslash S$ for which

$$\triangle y_i^k + \bar{c}_{ij}(x^k) \quad \text{with} \ \ i \in S$$

is minimal.

Now $\triangle y_i^k$ is the sum over all $\delta-$values computed until node i entered the set S . Thus for all nodes $i \in S$ the value $\triangle y_i^k$ is the length of the shortest path from s to i in $G(x^k)$ with respect to the costs $\bar{c}_{ij}(x^k)$ but this value is exactly $l(i)$ computed in the shortest augmenting path method.

Thus both methods compute the same minima and hence (modulo ties) the same $s\text{--}t\text{--}$path in $G(x^k)$, moreover the updating formulas for the dual variables are identical.

Comparing the two different formulas for calculating the minima it becomes clear why the shortest augmenting path technique is computationally superior in practice:

The minimum is computed over a much smaller set of candidates. More precisely, while in the primal–dual methods the candidates are the <u>edges</u> in the cut $(S, V \setminus S)$, the candidates for the shortest augmenting path method are the <u>nodes</u> in $\bar{S}$. Moreover one can see that the shortest augmenting path method comprises several dual updates of the primal dual algorithm into one dual update per phase.

8.2. Equivalence of the Network Simplex and the Shortest Augmenting Path Method

The simplex method leaves some options to the user concerning

- the initial feasible basic solution

- the choice of a proper "entering variable" rule and "leaving variable" rule.

In this section we analyse the special version with the following two specifications:

- "Big – M" start procedure and

- " most negative reduced cost rule" for entering nonbasic variables

for the min–cost flow problem.

Note that we are not imposing a specific "leaving edge" rule, thus we are not requiring strongly feasible trees here.

In the Big – M start procedure artificial edges (s, v) for all $v \in V \setminus \{s\}$ with sufficiently large cost $c_{sv} = M$ and infinite capacity u_{sv} are added and these edges form the initial basis–tree solution where the flow on each edge is zero except for the edge (s, t) which carries $z = b_t$ units of flow. The node s is taken as fixed root node for the basic trees.

Associated with any feasible tree is a basic feasible solution x which defines

113

an incremental graph $G(x)$. Now for an edge $(i,j) \notin E(T)$ let $c_{ij}(x)$ again denote the incremental cost. Here the incremental graph is always built with respect to the flow on the original edges only.

Let us define by $P_{i,j}$ the shortest augmenting path from i to j (i.e. directed i–j–path in $G(x)$) and by $T_{i,j}$ the unique path from i to j in the basis–tree and let $c(P_{i,j})$ resp. $c(T_{i,j})$ be the associated (incremental) cost. Then $c(T_{i,j}) \leq c(P_{i,j})$ holds since the path $T_{i,j}$ may be "blocked". Yet if $T_{i,j}$ is not blocked i.e.

$$x_e < u_e \quad \text{if } e \text{ is a forward edge in } T_{i,j}$$

$$x_e > 0 \quad \text{if } e \text{ is a reverse edge in } T_{i,j}$$

then equality holds.

Now the Big – M–network simplex method is started with the dual solution

$$y_s = 0$$

$$y_v = M \quad \text{for } v \in V \backslash \{s\} \; .$$

Since with this definition the reduced costs are given by

$$\bar{c}_{sj} = c_{sj} - M \quad \text{for all } (s,j) \in E \text{ with } M \text{ sufficiently large and}$$

$$\bar{c}_{ij} = c_{ij} \quad \text{for all other edges } (i,j) \text{ in } E$$

the first edge to enter the basis will be as in the shortest augmenting path method the "cheapest" edge incident with root s and the artificial edge incident with the other end node of the incoming edge will be pivoted out (by a degenerate pivot).

Now suppose that the simplex method has computed the shortest augmenting paths to a set S of permanently labeled nodes and has pivoted out the artifical edges incident with nodes in $S \backslash \{s\}$ then the current reduced cost of an edge (i,j) with $i \in S, \; j \in V \backslash S$ is

$$\bar{c}_{ij} = c(P_{s,i}) + c_{ij} - M.$$

The reduced costs of edges not contained in the cut $(S,\ V\backslash S)$ will not involve M. Hence the simplex–method will enter the edge which achieves

$$min\ \{c(P_{s,i}) + c_{ij}\ |\ i \in S,\ j \in V\backslash S\}$$

as the shortest augmenting path (or primal–dual) method does.

Thus the shortest (augmenting) s–t–path in G will be determined and then used for a first augmentation via a nondegenerate pivot.

In general the simplex basis will contain the edge (s, t) and artificial edges from s to a set of nodes, U say. The basis–tree T will consist of two subtrees, T_s with node set S rooted at s and T_t with node set $\bar{S} := V\backslash S$ rooted at t. These subtrees are connected by the edge (s, t).

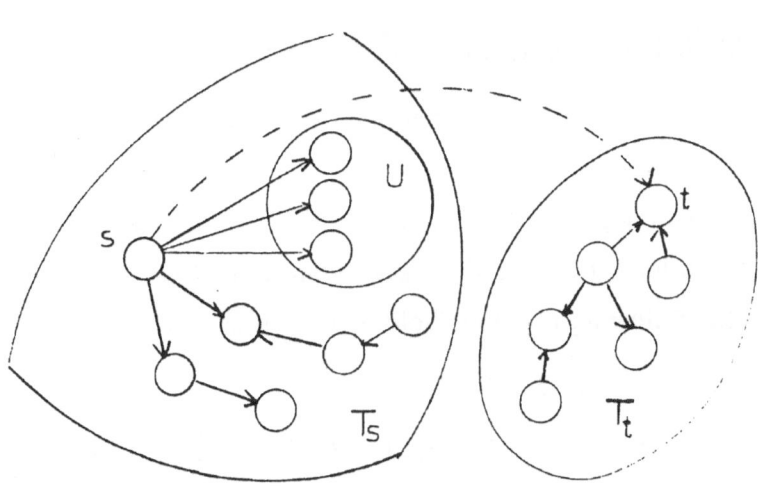

Figure 8.1. Example for a feasible basis

It suffices to give the prove for the case $U = \emptyset$ since a network yielding identical reduced costs may be obtained by replacing each artificial edge (s, u) for $u \in U$ by an (artificial) edge (u, t) with cost and capacity zero.

Now assume inductively that the simplex–method has augmented along a sequence of shortest augmenting paths and that

$$c(T_{s,j}) \leq c(P_{s,j}) \quad \text{for all } j \in S$$
$$c(T_{j,t}) \leq c(P_{j,t}) \quad \text{for all } j \in \bar{S} \ .$$

For $(i,j) \in (S, \bar{S})$ the reduced cost is given by

$$\bar{c}_{ij} = c(T_{s,i}) + c_{ij}(x) + c(T_{j,t}) - M$$

and under the "most negative reduced cost rule" the simplex method will enter the edge from $(S, \bar{S})$ which minimizes

$$c(T_{s,i}) + c_{ij}(x) + c(T_{j,t}) \ .$$

Now by inductive hypothesis

$$\min_{\substack{i \in S \\ j \in \bar{S}}} \{c(T_{s,i}) + c_{ij}(x) + c(T_{j,t})\} \leq \min_{\substack{i \in S \\ j \in \bar{S}}} \{c(P_{s,i}) + c_{ij}(x) + c(P_{j,t})\} = c(P^*)$$

where P^* is the shortest augmenting path connecting s and t .

Hence in the case of a nondegenerate pivot flow is sent along a shortest augmenting s–t–path again.

In any case the ensuing pivot (which may be degenerate) will generate new sets S' and $\bar{S}'$ and we have to show that with respect to the new trees and new reduced costs the following relations hold again:

$$c(T_{s,i}^{\text{new}}) \leq c(P_{s,i}^{\text{new}}) \quad \text{for all } i \in S'$$
$$c(T_{j,t}^{\text{new}}) \leq c(P_{j,t}^{\text{new}}) \quad \text{for all } j \in \bar{S}' \ .$$

Suppose without loss of generality that $S \backslash S' \neq \emptyset$, otherwise the following arguments apply to $\bar{S} \backslash \bar{S}' \neq \emptyset$.

Edmonds and Karp [1972] have shown that an augmentation of any amount (possibly zero) along a shortest augmenting path cannot reduce $c(P_{s,i})$ or $c(P_{i,t})$. Thus it suffices to show that

$$c(T_{j,t}^{\text{new}}) \leq c(P_{j,t}^{\text{old}}) \quad \text{for all } \ j \in S \backslash S'$$

Note that $c(T_{i,j}^{\text{new}}) = c(T_{i,j}^{\text{old}})$ unless the path in the new tree contains the incoming edge.

Now let $k \in S \backslash S'$. There is a path of edges connecting k to s in T_s^{old}. Let r be the first node on this path that is in the augmenting path P^*. Since P^* was a shortest augmenting path we know

$$c(T_{r,k}^{\text{old}}) + c(P_{k,t}^{\text{old}}) \geq c(T_{r,t}^{\text{new}})$$

From that we can conclude that

$$c(P_{k,t}^{\text{new}}) \geq c(P_{k,t}^{\text{old}}) \geq c(T_{r,t}^{\text{new}}) - c(T_{r,k}^{\text{old}})$$
$$= c(T_{r,t}^{\text{new}}) + c(T_{k,r}^{\text{new}}) = c(T_{r,t}^{\text{new}}).$$

Thus we have shown that the network simplex method with these two specifications performs (modulo ties) the same sequence of augmentations as the primal-dual approach or the shortest augmentation path method.

Here the "most negative reduced cost" rule for the entering edge is one of the classical rules (cf. Danzig [1963]). Variants of this rule —so called "altered most negative" rules— are used in most of the efficient network flow software packages. (cf. Bradley et al. [1977]; Mulvey [1978])

Mulvey [1972] has tested several options for choosing the initial basic solution (tree) and he found the Big – M–method to be highly superior over other approaches like Phase I/ Phase II starting routines etc.

Thus it is not only an interesting theoretical result that the network simplex method can be made to perform like the shortest augmentation path method by choosing the right specifications. Options which make the network simplex method mimic the shortest augmentation path method are among the most efficient choices. It is easy to see that the number of nondegenerates pivots for any network simplex method with Big – M start and most negative reduced cost rule is bounded by b_t . This is due to the fact that for each nondegenerate pivot the flow on the artifical edge (s.t) is decreased by an intregral amount $\Theta \geq 1$.

Recently Roohy-Loleh [1980] could show that if <u>strongly</u> feasible trees are maintained the length of a sequence of degenerate pivots is bounded by $|V| - 2$. This is due to the fact the most negative reduced cost rule will always enter an edge from the set $(S, \bar{S})$ i.e. an edge connecting the subtrees T_s and T_t. Then the special leaving edge rule will always choose an edge from the subtree T_t . Thus the nondegenerate pivot will decrease the cardinality of the set of nodes in T_t and hence no more than $|V| - 2$ consecutive degenerate pivots are possible.

This result shows that the total number of pivots is bounded by $(|V| - 2) \cdot b_t$. From this result follows immediately that the above method would need $O(n^2)$ pivot-operations for solving an $n \times n$ assignment problem i.e. is a polynomial algorithm for solving assignment problems.

PART IV

BIPARTITE MATCHING PROBLEMS

In this part we discuss two cornerstone problems in the class of general matching problems

- the assignment problem (AP) and

- the Hitchcock transportation problem (HTP)

We have shown already that both problems can be viewed as special MCFP's and each MCFP can be reduced to AP.

Thus the separate treatment and discussion of AP-algorithms and HTP-algorithms seems to be without justification and promise of benefit. Yet the reasons for analyzing algorithmic approaches for AP and HTP are the following:

- Historically the combinatorial structures and combinatorial (min-max) relations which are the basis of bipartite matching algorithms have been studied before network flow theory was developed.

- The structures to be handled here appear to be less "complex" than those to be handled in MCFP. But in principle AP and HTP contain already the whole complexity of network flow problems.

- The combinatorial structures to be handled are more "matching"-natured than those occuring in MCFP-methods and thus they built the proper basis for investigations on nonbipartite matching problems. In fact the approaches for AP and HTP will guide our development of the general (nonbipartite) matching theory and matching algorithms in nonbipartite graphs.

For the assignment problem we will discuss the following concepts

- the classical Hungarian method (Kuhn [1955])

- the primal approach of Balinski and Gomory [1964]

- the alternating basis simplex method (Barr et al. [1977])

- the shortest augmenting path method (Tomizawa [1972]).

There are quite a number of "other" algorithms presented in the literature for which one can show that they are equivalent to one of the above concepts.

In fact after discussing the above four main approaches we will again outline how the first three methods can be specialized to be interpretable as only different implementations of the shortest augmenting path principle.

For HTP we will not discuss the extensions of all four approaches, yet we will focus on the shortest augmenting path approach only. The primal HTP–approach was formulated by Balinski and Gomory [1964] together with the AP-approach. The alternating basis simplex method (which is essentially Cunningham's strongly feasible network method specialized for AP) has been extended to HTP by Barr et al. [1978] while Kuhn's Hungarian method was extended to HTP by Ford and Fulkerson [1957] (with the basic RAND–Report appearing in 1955 already).

As a main result for HTP we present the *scaling-method* of Edmonds and Karp [1972] which is the so far only "good" (combinatorial) algorithm known for solving HTP. We will analyse and introduce this approach from a combinatorial point of view as a special implementation of the shortest augmenting path approach.

Before we start with our discussion of the assignment problem we will introduce the bipartite cardinality matching problem. This problem occurs as a special subproblem in the primal–dual approach to AP and in this sense this problem plays the same role for AP as MFP did for MCMF.

Chapter 9. The Cardinality Matching Problem in Bipartite Graphs

Given a graph $G = (V, E)$ a *1-matching* (or simply *matching*) in G is a subset $M \subseteq E$ such that each vertex $v \in V$ is incident with at most one edge from M . By $\mathcal{M}$ we denote the set of all matchings in G , where $\mathcal{M} \neq \emptyset$ since $\emptyset \in \mathcal{M}$ holds. By $|M|$ we denote the cardinality of M . Then the *maximum cardinality matching problem* (CMP) is to find a matching in G the cardinality of which is as large as possible i.e.

$$max \ \{|M| \mid M \in \mathcal{M}\}. \tag{CMP}$$

An optimal solution M of CMP is called a *maximum cardinality matching*. A matching $M \in \mathcal{M}$ is called *perfect* iff every node $v \in V$ is incident to exactly one edge from M . If $G = (V_s, V_t, E)$ is a bipartite graph then a perfect matching is also called an *assignment*. Obviously any perfect matching solves CMP. In general we will denote by $\nu(G)$ the *matching number* of G i.e. the cardinality of a maximum cardinality matching.

Given a matching M in $G = (V, E)$ we denote by $V(M) \subseteq V$ the set of *M-saturated* nodes, i.e. those nodes which are incident with an edge from M . The set $U(M) := V \setminus V(M)$ is called the set of *M-unsaturated* or *isolated* nodes.

As already shown, CMP can be formulated as integer program.

$$max \quad x(E) \quad \text{subject to} \tag{CMP}$$
$$x(\delta(v)) \leq 1 \text{ for } v \in V$$
$$x \in \{0,1\}^E$$

or equivalently

$$max \quad 1'x$$
$$Ax \leq 1$$
$$x \in \{0,1\}^E$$

with A the node-edge incidence matrix of G . If G is bipartite the integrality conditions are superflous and CMP becomes a simple LP. Yet we will not apply

LP–theory to CMP but develop the purely combinatorial approaches for solving CMP.

9.1. Combinatorial Structures

In this section we focus on the bipartite case. Yet most of the definitions and theorems stated below are valid and useful for the general case, too. Hence we will explicitly state when the validity of a theorem is restricted to the bipartite case only.

A simple path $P = (V(P), E(P))$ resp. simple cycle $C = (V(C), E(C))$ in G is called M-alternating if its edges are alternately in M and not in M. An alternating path P is called M-augmenting if it connects two M–unsaturated nodes. If M is a matching and $P = (V(P), E(P))$ is an augmenting path (resp. alternating cycle) then $M \oplus P := (M \setminus E(P)) \cup (E(P) \setminus M)$ is again a matching and $|M \oplus P| = |M| + 1$ (resp. $|M \oplus P| = |M|$) holds.

Theorem 9.1.

If M and N are two matchings in G, a connected component of the graph $G' = (V, M \oplus N)$ is one of the following types:

(i)   *a single (isolated) node,*

(ii) *a cycle the edges of which are alternately in M and N,*

(iii) *a simple path the edges of which are alternately in M and N and the endnodes of which are unsaturated for one of the two matchings, M or N.*

Proof.   For any $v \in V$ we have $d_{G'}(v) \in \{0, 1, 2\}$. From this property the characterization of the components of G follows immediately. $\quad\square$

Let $K_i = (V_i, E_i), i = 1,\ldots,r$ be the connected components of G' which are of type (ii) and (iii), then $M = N \oplus K_1 \oplus \ldots \oplus K_r$ holds. From this the following relations are easy to see.

Corollary 9.2.   *(Berge [1957])*
M is a maximum cardinality matching in G iff G does not contain any M–augmenting path.

Corollary 9.3.
Let M and N be matchings in G with $|N| - |M| = r > 0$. Then $G' = (V, M \oplus N)$ contains at least r components which are (node–disjoint) augmenting paths relative to M .

Corollary 9.2 motivates the following approach for solving CMP.

<u>Procedure 9.1</u>   **Alternating path approach for solving CMP**

<u>Step 0.</u>   Determine $M \in \mathcal{M}$.

<u>Step 1.</u>   Check for existence of an M–augmenting path.

If no such path exists: Stop, M is optimal.

Otherwise with P an M–augmenting path, goto <u>Step 2.</u>

<u>Step 2.</u>   Set $M := M \oplus P$ and goto <u>Step 1</u> .

In fact all (efficient) algorithms for solving CMP follow this scheme in the bipartite case as well as in the general case. We will call one iteration of this procedure, i.e. the check for existence of an augmenting path and the augmentation, a *phase* of the procedure. Then it is obvious that at most $\frac{1}{2} \cdot |V|$ phases are needed to compute a maximum cardinality matching.

123

The remainig problems are

– to find augmenting paths relative to a matching resp.

– to verify that no such augmenting path exists.

Here we will have to apply different tools in the bipartite case and the general case. The complexity of the complete procedure can then be calculated as $|V|$ times the complexity for solving the phase-subproblems.

From now on let $G = (V_s, V_t, E)$ be a <u>bipartite graph</u> with bipartition $V = V_s \cup V_t$, i.e. for every edge $\{i, j\} \in E$ with $i \in V_s$ we have $j \in V_t$.

Any CMP on a bipartite graph can easily be transformed into a MFP on a related digraph. For that purpose we add a source $s \notin V$ and a sink $t \notin V$ and introduce directed edges (s, i) for $i \in V_s$ and (j, t) for $j \in V_t$ each having capacity $u_{si} = u_{jt} = 1$. All edges $\{i, j\} \in E$ with $i \in V_s$ are directed from i to j, i.e. they are interpreted as directed edges (i, j) , and their capacities are set to infinity.

Then any feasible (integer valued) s-t-flow x in this network induces a matching M in G via the relation

$$\{i, j\} \in M \Leftrightarrow x_{ij} > 0$$

with the property $|M| = z(x)$, i.e. the cardinality of M equals the flow value of x .

Hence any integer max s-t-flow induces a maximum cardinality matching and $\nu(G)$ can be characterized by the max-flow-min-cut relation.

Yet we will not treat CMP in bipartite graphs as a special case of MFP. We prefer a more matching natured approach because of the reasons given in the introduction.

With respect to a matching M in a bipartite graph $G = (V_s, V_t, E)$ we denote by $V_s(M)$ resp. $V_t(M)$ the nodes in V_s resp. V_t which are saturated

124

by M. Our first problem is then to characterize bipartite graphs which allow perfect matchings resp. assignments.

Obviously $G = (V_s, V_t, E)$ does not allow perfect matchings (assignments) if $|V_s| \neq |V_t|$. A slightly more general question is the question under which conditions there exists a matching M which saturates all nodes in V_s.

Theorem 9.4.   *(Hall [1935])*
$G = (V_s, V_t, E)$ *allows matchings which saturate* V_s *iff* $|N(S)| \geq |S|$ *for all* $S \subseteq V_s$.

This theorem is not a "good" characterization of bipartite graphs allowing V_s-saturating matchings since the number of conditions to be verified is growing exponentially with the cardinality of V_s.

The following theorem which is even older than Hall's theorem gives a "nice" characterization via a combinatorial *min-max-relation*, and this relation will be used as an optimality condition resp. stopping–criterion in the augmenting–path approach for solving CMP.

A (node–) *cover* of G is a subset $K \subseteq V$ such that for all $\{i, j\} \in E$ we have $\{i, j\} \cap K \neq \emptyset$. Let $\mathcal{K}$ be the set of all covers of G then $\tau(G) := min\{|K| \mid K \in \mathcal{K}\}$ is called the *cover-number* of G.

It is easy to see that $\nu(G) \leq \tau(G)$ holds for all (not necessarily bipartite) graphs $G = (V, E)$. In fact given a matching M in G for every edge $\{i, j\} \in M$ at least one of the two nodes i and j has to be included in any cover.

Now for bipartite graphs equality of both numbers can be shown:

Theorem 9.5.   *(König [1936])*
In a bipartite graph $G = (V_s, V_t, E)$ *the maximal cardinality of a matching equals the minimal cardinality of a cover i.e.* $\tau(G) = \nu(G)$ *holds.*

Proof. Let $M^* \in \mathcal{M}$ with $|M^*| = \nu(G)$ and $U_s = V_s \setminus V_s(M^*)$. If $U_s = \emptyset$ then $K = V_s$ is a cover with $|K| = |M^*|$. Otherwise define $Z \subseteq V$ to be the set of nodes which are "reachable" from a node $u \in U_s$ by an M^*–alternating path in G. Let $S := Z \cap V_s$ and $T := Z \cap V_t$. Then it is easy to see that

(i) all nodes $t \in T$ are M^*-saturated, since otherwise G would allow a M^*-augmenting path,

(ii) $N(S) = T$.

Now define $K = (V_s \setminus S) \cup T$ then K is a cover in G and $|K| = |M^*|$ holds.

$\square$

The above proof of König's theorem uses in a way the nonexistence of an M^*-augmenting path in a constructive manner. Another elegant proof is due to Lovasz [1975]. In fact König's theorem can also be proved using LP-duality or the max-flow-min-cut theorem of Ford–Fulkerson. Moreover König's theorem can be used to prove Hall's theorem. In a CMP-algorithm for bipartite graphs König's theorem can be used to show the optimality of the solution. Egervary [1931] has proven an equivalent relation for $0 - 1$–matrices, therefore the theorem is also called the König–Egervary theorem.

The following definiton describes a structure which is basic for any algorithmic treatment of matchings in graphs. Again this definition is valid for bipartite as well as general, not bipartite graphs.

Let M be a matching in $G = (V, E)$ and $T = (V(T), E(T))$ a tree in G. Then T is called an *alternating tree* with respect to M (*M–alternating tree*) iff:

(i) $V(T)$ contains exactly one unsaturated node, u say, the so-called *root* of T.

(ii) Each path from $v \in V(T)$ to u in T is an M-alternating path.

(iii) $\delta(V(T)) \cap M = \emptyset$, i.e. T contains all edges from M which are incident with nodes in $V(T)$.

The nodes in $V(T)$ for which the unique path to u has an even number of edges are called *outer* nodes of T and the set of outer nodes is denoted by $O(T)$; the other nodes in $V(T)$ are called *inner* nodes and the set of inner nodes is denoted by $I(T)$. Note that if T is an alternating tree in a bipartite graph $G = (V_s, V_t, E)$ with $u \in V_s$ then $O(T) \subseteq V_s$ and $I(T) \subseteq V_t$ holds.

Alternating trees can be used to find augmenting paths with respect to a matching M. For that purpose one starts from an isolated node $u \in V$ and makes it the root of the tree T which contains only this node and no edges at all. This is obviously an M-alternating tree with $O(T) = \{u\}$.

Now let $T = (V(T), E(T))$ be an M-alternating tree, $v \in O(T)$ and $\{v, w\} \in E \setminus E(T)$, then we can distinguish the following four cases:

Case 1: w is isolated with respect to M.

In this case the alternating path from v to u in T together with the edge $\{v, w\}$ forms an augmenting path with respect to M.

Case 2: w is "matched" with node $x \notin V(T)$.

In this case we can "grow" the tree T by appending the edges $\{v, w\}$ and $\{w, x\}$ i.e.

$$V(T) := V(T) \cup \{w, x\}$$
$$E(T) := E(T) \cup \{\{v, w\}, \{w, x\}\}$$

Case 3: $w \in I(T)$

In this case the edge $\{v, w\}$ gives no new information which can be used for finding an augmenting path.

Case 4: $w \in O(T)$

This case is possible only if G is not bipartite. We will discuss the implications of this case in the section on general 1–matchings.

An alternating tree $T = (V(T), E(T))$ for which only <u>case 3</u> occurs, i.e. $N(O(T)) \subseteq I(T)$ holds, is called a *Hungarian tree*, and this property guarantees that no M–augmenting path with endnode u exists. The following lemma holds again in bipartite and general graphs.

Lemma 9.6.

Let $T = ((V(T), E(T))$ be a Hungarian tree with root u . Then the following properties hold:

(i)   There does not exist an M-augmenting path with endnode u .

(ii)   When "growing" another alternating tree from another isolated node u' all edges in $\delta(V(T))$ need not be examined again.

Proof.   ad i): Assume an M–augmenting path P with endnode u and let w be the first node in P when starting from u which is not in $V(T)$ and let $v \in V(T)$ be the predecessor of w in P . Now $\{v, w\} \in M$ contradicts the fact that T is an alternating tree. On the other hand $\{v, w\} \notin M$ implies $v \in O(T)$, thus $N(O(T)) \not\subseteq I(T)$ and hence T is not a Hungarian tree.

ad ii): Let $\overline{M}$ be a matching in G with $\overline{M} \cap E(T) = M \cap E(T)$ and assume that $P = (V(P), E(P))$ is an $\overline{M}$-augmenting path with $V(P) \cap V(T) \neq \emptyset$. Let v_1 and v_2 be two vertices in $V(T) \cap V(P)$ such that the partial path of P from v_1 to v_2 is contained in T . Then $\{v_1, v_2\} \subseteq I(T)$ since T is a Hungarian tree, but then it is easy to see that P cannot be an $\overline{M}$–alternating path.

$\square$

Now let $G = (V_s, V_t, E)$ be a bipartite graph, then starting from an isolated node $u \in V_s$ as root after at most $|V_s|$ "grow–steps" either an augmenting path is detected or the alternating tree becomes a Hungarian tree. Thus the phase-subproblem in the bipartite case can be solved by successively growing alternating trees from all isolated nodes. After at most $|V_s|$ phases a maximum cardinality matching, M^* say, has been constructed. If $|M^*| = |V_s|$ then M^*

is obviously a maximum cardinality matching. Otherwise the optimality of M^* can be shown using König's theorem :

Let K be the set of all nodes in V_t which are contained in any Hungarian tree plus the set of nodes in V_s which are not contained in any Hungarian tree. Then K is a cover of G with $|K| = |M^*|$ and thus M^* is of maximal possible cardinality.

A variant of this "alternating tree approach" is to grow alternating trees simultaneously from all isolated nodes in V_s . Then instead of a single alternating tree we have to handle a collection of alternating trees, a so-callled *alternating forest*. With $F = (V(F), E(F))$ an alternating forest we denote by $O(F)$ resp. $I(F)$ the set of outer resp. inner nodes in the alternating trees. Then F is called *Hungarian forest* if $N(O(F)) \subseteq I(F)$ holds.

9.2. Augmenting Path Methods

In the following we state a simple labeling-method for constructing augmenting paths resp. for showing the optimality of a matching. Here we will follow the "alternating–forest–approach" and built up alternating trees from all unsaturated nodes $s \in V_s$ simultaneously. For that purpose let $M \in \mathcal{M}$. During the course of the algorithm nodes in $V_s \cup V_t$ may be either *unlabeled* or *labeled*. At the beginning all nodes are unlabeled.

Procedure 9.2.   **Alternating-forest-CMP-labeling method**

Step 0:   Label all nodes $u \in V_s \setminus V_s(M)$.

Step 1:   Choose unscanned but labeled node $s \in V_s$.

   If no such node exists, STOP: M is of maximal cardinality.

   Otherwise, given such a node s, goto Step 2.

<u>Step 2:</u>   Scan node s in the following way:

For all $t \in N(s)$ unlabeled assign label $p(t) := s$.

If $t \in U(M)$, STOP: there exists an augmenting path

connecting t with a node $u \in U(M) \cap V_s$.

Otherwise for $\{s', t\} \in M$, label s' .

Goto <u>Step 1.</u>

(Note that the augmenting path found in Step 2, can easily be traced back using the predecessor label p(.))

The above labeling method requires $O(|E|)$ elementary operations and this includes the amount for augmenting along the augmenting path found in <u>Step 2.</u> Since at most $|V_s|$ phase–subproblems have to be solved we end up with a method of complexity $O(|V_s| \cdot |E|)$ for solving the cardinality matching problem.

The validity of this labeling method can again be shown applying König's theorem.

A variant of this labeling method would be to grow the alternating trees until no more nodes can be labeled and then to augment along several node–disjoint augmenting paths *simultaneously*. In fact this simple idea combined with the concept of "shortest augmenting paths" has lead to an improvement with respect to computational complexity. This was first observed and implemented by Hopcroft and Karp [1973].

The following definitions and theorems are valid again for the bipartite matching problem as well as for the CMP on general non–bipartite graphs.

Let M be a matching then the *length* of an augmenting path P is $|E(P)|$, the number of edges in P . An augmenting path P is called *shortest* with respect to M if $|E(P)|$ is of least cardinality among all augmenting paths relative to M .

Theorem 9.7.

Let M be a matching, P a shortest augmenting path relative to M and P' an augmenting path relative to $M \oplus P$. Then

$$|E(P')| \geq |E(P)| + |E(P) \cap E(P')| \, .$$

Proof.   (cf. Hopcroft and. Karp [1973])

Now assume that starting from $M_0 = \emptyset$ the labeling method has computed a sequence $M_0, M_1, M_2, \ldots, M_i, \ldots$ of matchings where $M_{i+1} = M_i \oplus P_i$ with P_i a shortest augmenting path relative to M_i . Then the following properties are a consequence of the above theorem:

Property 1 : $|E(P_i)| \leq |E(P_{i+1})|$ for $i = 0, 1, \ldots$.

Property 2 : $|E(P_i)| = |E(P_j)|$ implies $V(P_i) \cap V(P_j) = \emptyset$.

Moreover the following theorem holds:

Theorem 9.8.   *(Hopcroft and Karp [1973])*
The number of distinct integers in the sequence

$$|E(P_0)| \, , |E(P_1)| \, , \ldots, |E(P_i)| \, , \ldots$$

is less than or equal to $2 \lfloor \sqrt{\nu(G)} \rfloor + 2$.

Thus the computation of the sequence $M_0, M_1, \ldots, M_i, \ldots$ breaks into at most $2 \lfloor \sqrt{\nu(G)} \rfloor + 2$ phases within each of which all the augmenting paths found are node-disjoint and of the same length. Since the paths are node–disjoint they are all augmenting paths relative to the matching with which the phase is begun. This property induces an alternative way of computing a maximum cardinality matching:

<u>Procedure 9.3.</u>   **Hopcroft–Karp–approach for solving CMP**

<u>Step 0:</u>   Set $M := \emptyset$.

<u>Step 1:</u>   Let $l(M)$ be the length of a shortest augmenting path relative to M . Find a maximal set of augmenting paths $\{P_1, \ldots, P_t\}$ relative to M with the property that

(i) $|E(P_i)| = l(M)$ for $1 \leq i \leq t$ and

(ii) $V(P_i) \cap V(P_j) = \emptyset$ for $1 \leq i, j \leq t$.

If no augmenting path exists, STOP: M is of maximal cardinality.

Otherwise given $P_1, \ldots, P_t$ goto <u>Step 2</u>.

<u>Step 2:</u>   Set $M := M \oplus P_1 \oplus P_2 \oplus \cdot \oplus P_t$ and goto <u>Step 1</u>.

(Note, that we say that a set is maximal with a given property if it has the property and is not properly contained in any set that also has the property.) While the basic augmenting path approach needs $O(|V|)$ phases the Hopcroft-Karp-approach needs $O(\sqrt{|V|})$ only. Yet the modified phase-subproblem is to find a maximal set of node-disjoint augmenting paths instead of just any augmenting path. This seems to be a more complex problem but Hopcroft and Karp [1973] could give a labeling method which in the case of a bipartite graph solves this subproblem in $O(|E|)$ steps, too, which leads to an algorithm for solving CMP in bipartite graphs of complexity $O(\sqrt{|V|} \cdot |E|)$. (Some years later Even and Kariv [1975], Micali and Vazirany [1980] extended this idea successfully to the case of non–bipartite graphs, cf. chapter 12)

For the bipartite case Darby–Dowman [1981] has implemented both approaches and has run tests on a significant number and set of problems. His results indicate that in practice the Hopcroft–Karp–approach is inferior to the basic approach when the single augmentations are always performed along shortest augmenting paths. This is due to the fact that the amount of work for solving the modified subproblem is (in practice) larger than the amount for several single augmentations. These results came true in our experiments with different CMP–procedures, too.

Chapter 10. The Assignment Problem

In this chapter we discuss the first key problem within the class of general matching problems. The reasons for extensively analyzing algorithmic approaches to the assignment problem have been stated in the introduction of this part. Yet we feel that in addition the following facts justify our approach and should be mentioned here:

- The primal–dual algorithm for solving AP — also known as Hungarian method — was the pioneer of all network flow and matching algorithms and it was the motivation of many successful approaches to more general network and matching problems.

- Certain refinements of the strongly feasible network simplex method applied to AP lead to a "good", i.e. a polynomial algorithm. To our knowledge AP is the so far only "class" of problem for which a polynomial simplex-type method has been developed. And again the results and insights obtained for AP may lead to more successes of this kind for more general network flow or matching problems.

Given a bipartite graph $G = (V_s, V_t, E)$ and $\mathcal{M}_p \subseteq \mathcal{M}$ the set of perfect matchings in G, the assignment problem can be formulated in the following way

$$min \ \{c(M) \mid M \in \mathcal{M}_p\} \qquad \text{(AP)}$$

where for each subset $F \subseteq E$ we define $c(F) = \sum_{l \in F} c_l$. Another formulation using the node–edge incidence matrix A of G is

$$min \ \{c'x \mid Ax = 1, x \geq 0 \text{ integer valued}\}. \qquad \text{(AP)}$$

A necessary conditon for G to allow perfect matchings is $|V_s| = |V_t| (=: n)$. Introducing artifical edges with sufficiently large cost if necessary we may assume that G is a *complete* bipartite graph and this problem is denoted by AP_n then.

In this case we can identify both sets V_s and V_t with the set $\{1, \ldots, n\}$ and formulate AP_n as a problem of finding an optimal permutation. For that purpose let S be the set of all permutations (bijective mappings) $\varphi : N \rightarrow N$, then AP_n is equivalently formulated as

$$min \ \{\sum_{i=1}^{n} c_{i,\varphi(i)} \mid \varphi \in S\}.$$

In the literature the assignment problem is often treated as a maximization problem (cf. Thompson [1981]) and instead of perfect matchings one may also like to find the (not necessarily perfect) matching of maximum weight in G (cf. Lawler [1976]). Indeed it can be shown that all these problems can be transformed into an equivalent AP_n .

In the following we discuss some well-known important features of AP_n .

10.1. The Assignment Polytope

The $n \times n$ assignment polytope P_n is the set $x = (x_{ij})$ of "vectors" fulfilling

$$\sum_{j=1}^{n} x_{ij} = 1 \quad \text{for } i = 1, \ldots, n$$

$$\sum_{i=1}^{n} x_{ij} = 1 \quad \text{for } j = 1, \ldots, n$$

$$x_{ij} \geq 0 \quad \text{for } 1 \leq i, j \leq n$$

or equivalently

$$P_n = \{x \mid Ax = 1, x \geq 0\}$$

with A the $2n \times n^2$ node edge incidence matrix of the complete bipartite graph $K_{n,n}$.

The assignment polytope has been studied extensively in the literature and an excellent expository account describing interesting properties of P_n is Balinski and Russakoff [1974].

The most important and useful property is the following:

Theorem 10.1. *(Birkhoff [1946])*

The extreme points $x \in P_n$ thought of as $n \times n$ matrices are precisely the permutation matrices of order n.

Thus the assignment problem AP_n can be formulated and solved as a linear program

$$min \; \{c'x \mid x \in P_n\}.$$

A different integrality-proof for P_n is based on the theorem of Hoffman and Kruskal [1956] which shows the total unimodularity of A .

A basis B of P_n, i.e. a maximal linear independent set of columns of A , has cardinality $2n - 1$. Hence in any feasible basic solution exactly n of the $2n - 1$ basic variables have value 1 and $n - 1$ basic variables have value 0 . Thus AP_n thought of as a linear program is extremely degenerate.

Now given a basic feasible solution $x \in P_n$ with associated basis $B(x)$ the 1–valued variables induce an assignment $M(x)$. The subgraph $T = (V, B(x))$ induced by the basis $B(x)$ contains no cycle, since otherwise $B(x)$ is not linear independent, and it spans G , since $M(x) \subseteq B(x)$ spans G . Thus $T = (V, B(x))$ is a spanning tree of G . The following theorem comprises some of the relations among extreme points and bases of P_n .

Theorem 10.2.

(i) *The extreme points of P_n are in one-to-one correspondence with the assignments in $K_{n,n}$.*

(ii) *The bases (feasible bases) B of P_n are in one-to-one correspondence with the spanning trees of $K_{n,n}$ (the spanning trees of $K_{n,n}$ which contain a perfect matching of $K_{n,n}$) .*

(iii) *Two feasible bases B_1, B_2 of P_n are neighbours iff $G_{1,2} := (V, B_1 \cup B_2)$ contains exactly one cycle.*

(iv)   Two extreme points x_1, x_2 of P_n are neighbours iff

$$G_{1,2} := (V, M(x_1) \cup M(x_2))$$

contains exactly one circuit.

Proof.   (cf. Balinski and Russakoff [1974].)

From the above theorem and "Cayleys–formula" for the number of spanning trees it follows that to each extreme point of P_n there correspond $2^{n-1}n^{n-2}$ feasible bases. Thus P_n has $2^{n-1}n^{n-2}n!$ feasible bases at all.

A highly interesting result obtained by Balinski and Russakoff, too, is the fact that the Hirsch conjecture is true for P_n .

Theorem 10.3.
Given any pair x, y of feasible bases of P_n , then y can be obtained from x by a sequence of at most $2n - 1$ pivot operations.

Thus in theory any AP_n can be solved in at most $2n - 1$ simplex iterations. Yet in practice we never know the right $2n - 1$ pivots, unfortunately.

As for CMP, augmenting paths and alternating cycles are a central concept in AP–aproaches, too. Yet in weighted graphs we will associate a different cost or length–function with such paths and cycles.

Given a matching M in $G = (V, E)$ and $P = ((V(P), E(P))$ a M–augmenting path resp. M–alternating cycle in G we define

$$c(P) := c(E(P) \setminus M) - c((E(P) \cap M)$$

the "length" or "cost" of P . With this definition the relation

$$c(M \oplus P) = c(M) + c(P)$$

holds.

10.2. The Hungarian Method

Historically the Hungarian method developed by Kuhn [1955] was motivated by purely combinatorial arguments — more precisely the max–matching–
min–cover relation for bipartite graphs. Kuhn has credited the importance of the
König–Egervary theorem for his method by suggesting the name "Hungarian"-
method. Later this approach was adapted to other problems where combinatorial max–min–relations could be shown and it also motivated the development
of the primal–dual method for solving general linear programs.

In this section we will introduce the Hungarian method as a specialization
of the general primal-dual LP-approach and we will exploit the combinatorial
nature of the respective primal and dual subproblems.

The assignment problem in a bipartite graph $G = (V_s, V_t, E)$ with cost
function $c : E \rightarrow \mathbb{R}_+$ can be formulated as a linear program

$$min \sum_{i \in V_s, j \in V_t} c_{ij} \cdot x_{ij} \quad \text{subject to} \qquad \text{(AP)}$$

$$\sum_{j \in N(i)} x_{ij} = 1 \quad \text{for } i \in V_s$$

$$\sum_{i \in N(j)} x_{ij} = 1 \quad \text{for } j \in V_t$$

$$x_{ij} \geq 0 \quad \text{for } i \in V_s, j \in V_t.$$

The dual of the assignment problem is given by

$$max \sum_{i \in V_s} u_i + \sum_{j \in V_t} v_j \qquad \text{(DAP)}$$

$$u_i + v_j \leq c_{ij} \quad \text{for } \{i, j\} \in E$$

$$u_i, v_j \quad \text{unrestricted in sign}$$

As for MCFP we call $\bar{c}_{ij} := c_{ij} - u_i - v_j$ the reduced cost of edge $\{i, j\}$. Now
let $(u, v) \in \mathbb{R}^{2n}$ be a dual feasible vector then define

$$E(u, v) = \{\{i, j\} \in E \mid \bar{c}_{ij} = 0\}$$

the set of so-called admissible edges with respect to (u, v). With this definition the restricted primal problem (RAP) reads:

$$min \ \xi = \sum_{i \in V_s} d_i + \sum_{j \in V_t} e_j \qquad \text{(RAP)}$$

$$\sum_{j \in N(i)} x_{ij} + d_i = 1 \quad \text{for } i \in V_s$$

$$\sum_{i \in N(j)} x_{ij} + e_j = 1 \quad \text{for } j \in V_t$$

$$x_{ij} \geq 0 \quad \text{for } \{i, j\} \in E$$

$$x_{ij} = 0 \quad \text{for } \{i, j\} \in E \setminus E(u, v)$$

$$d_i \geq 0 \quad \text{for } i \in V_s$$

$$e_j \geq 0 \quad \text{for } j \in V_t.$$

An easy calculation gives

$$\xi = \sum_{i \in V_s} (1 - \sum_{j \in N(i)} x_{ij}) + \sum_{j \in V_t} (1 - \sum_{i \in N(j)} x_{ij})$$

$$= |V_s| + |V_t| - 2 \sum_{\{i,j\} \in E} x_{ij}$$

$$= 2(n - \sum_{\{i,j\} \in E} x_{ij}) \quad \text{with } n := |V_s| = |V_t|.$$

This shows that RAP is equivalent to the CMP on the subgraph $G(u, v) = (V_s, V_t, E(u, v))$. Thus RAP can be solved by one of the CMP–labeling methods introduced in the last chapter. Let us denote by M the maximum cardinality matching in $G(u, v)$ (optimal solution for RAP) produced by this labeling method. If $|M| = n$ then $\xi = 0$ and M is an optimal solution for AP, too. Otherwise the associated optimal value of RAP is $\xi_{OPT} = 2(n - |M|) > 0$ and we have to consider the dual of RAP i.e.

$$max \sum_{i \in V_s} u_i + \sum_{j \in V_t} v_j \qquad \text{(DRAP)}$$

$$u_i + v_j \leq 0 \quad \text{for } \{i, j\} \in E(u, v)$$

$$u_i \leq 1 \quad \text{for } i \in V_s$$

$$v_j \leq 1 \quad \text{for } j \in V_t.$$

Since M is not perfect, the labeling method has stopped with a Hungarian forest. With respect to the final labeling, let S^* and T^* be the nodes in V_s and V_t, respectively, which are labeled, i.e., contained in a Hungarian tree.

From the results in section 9.1 we know that $K = (V_s \setminus S^*) \cup T^*$ is a cover of $G(u, v)$ and $|K| = |M|$ holds.

Theorem 10.4.

Let

$$\bar{u}_i = \begin{cases} +1, & \text{for } i \in S^*, \\ -1, & \text{for } i \in V_s \setminus S^*. \end{cases}$$

$$\bar{v}_j = \begin{cases} -1, & \text{for } j \in T^*, \\ +1, & \text{for } j \in V_t \setminus T^*. \end{cases}$$

Then $(\bar{u}, \bar{v})$ is an optimal solution for DRAP.

Proof.

(i) $(\bar{u}, \bar{v})$ is feasible for DRAP since $\bar{u}_i + \bar{v}_j \in \{-2, 0, 2, \}$ for $i \in V_s, j \in V_t$; and $\bar{u}_i + \bar{v}_j > 0 \Leftrightarrow i \in S^*$ and $j \in V_t \setminus T^*$. Thus $\{i, j\} \in E(u, v) \Rightarrow \bar{u}_i + \bar{v}_j \leq 0$.

(ii) $(\bar{u}, \bar{v})$ is an optimal solution for DRAP since

$$\sum_{i \in V_s} \bar{u}_i + \sum_{j \in V_t} \bar{v}_j = 2(|V_s| - |V_s \setminus S^*| - |T^*|)$$

$$= 2(n - |K|) = 2(n - |M|) = \xi_{OPT}.$$

$\square$

140

Now $\xi_{OPT} > 0$ and we have to distinguish two cases:

<u>Case 1</u>: $\bar{u}_i + \bar{v}_j \leq 0$ for all $\{i,j\} \in E \setminus E(u,v)$.

In that case the primal problem is infeasible, i.e. $G = (V_s, V_t, E)$ does not contain an assignment.

If we assume that G contains perfect matchings, only the second case is possible:

<u>Case 2</u>: $\bar{u}_i + \bar{v}_j > 0$ for at least one $\{i,j\} \in E \setminus E(u,v)$.

In this case we have to calculate

$$\Theta = \min \{\bar{c}_{ij}/(\bar{u}_i + \bar{v}_j) \mid \bar{u}_i + \bar{v}_j > 0, \{i,j\} \notin E(u,v)\}.$$

Now we know

$$\bar{u}_i + \bar{v}_j > 0 \quad \Leftrightarrow \quad i \in S^* \quad \text{and } j \in V_t \setminus T^*$$
$$\Leftrightarrow \quad \bar{u}_i + \bar{v}_j = 2$$

and thus

$$\Theta = \min \{\bar{c}_{ij}/2 \mid i \in S^*, j \in V_t \setminus T^*\}.$$

Now the "improved" dual solution $(\tilde{u}, \tilde{v})$ is given by the following formula ("dual-change")

$$\tilde{u}_i := \begin{cases} u_i + \delta, & \text{if } i \in S^*, \\ u_i, & \text{if } i \in V_s \setminus S^* \end{cases}$$

$$\tilde{v}_j := \begin{cases} v_j - \delta, & \text{if } j \in T^*, \\ v_j, & \text{if } j \in V_t \setminus T^* \end{cases}$$

where $\delta := 2\Theta$.

From the general primal-dual theory we know:

(i) $\quad \{i,j\} \in M \quad \Rightarrow \quad c_{ij} - \tilde{u}_i - \tilde{v}_j = 0$, thus $\{i,j\} \in E(\tilde{u}, \tilde{v})$

and

(ii) $K = V_S \setminus S^* \cup T^*$ is not a cover of $G(\tilde{u}, \tilde{v})$ since at least one edge $\{i, j\}$ with $i \in S^*$ and $j \in V_t \setminus T^*$ fulfills $c_{ij} - \tilde{u}_i - \tilde{v}_j = 0$, i.e. $\{i, j\} \in E(\tilde{u}, \tilde{v})$.

In addition to these two general properties of the primal-dual approach one can show that after the dual change in the Hungarian method all edges contained in the Hungarian forest in $G(u, v)$ will again become admissible with respect to $(\tilde{u}, \tilde{v})$.

Note that the optimal solution of DRAP is not unique and that different optimal vectors $(\tilde{u}, \tilde{v})$ lead to different updating formulas (cf. Papadimitriou and Steiglitz [1982]).

As demonstrated above the primal–dual method boils down to a purely combinatorial algorithm. RAP as well as DRAP can be solved by means of the CMP labeling method. Thus one can formulate this method as a labeling procedure, too, which works on the graph $G = (V_s, V_t, E)$ directly.

In the primal phase alternating trees are grown in certain subgraphs $G(u, v)$ of G and in the dual phase the minimal reduced cost of an "uncovered" edge in G is used to update the dual solution.

Thus after each dual change at least one alternating tree can be grown which leads to either an augmentation or at least two newly labeled nodes. Since after at most n dual updates an augmentation must occur and at most n augmentations are needed the procedure stops with an optimal assignment after at most n^2 dual changes resp. iterations. Here the solution of the primal subproblem, the determination of Θ and the dual change can be performed in $O(n^2)$ steps. Thus a rough estimation for the complexity of the primal-dual method is $O(n^4)$.

Yet more clever updating techniques have been developed which reduce the total compexity to $O(n^3)$ (cf. Lawler [1976], Papadimitriou and Steiglitz [1982]). We will present such an improvement later, when we discuss the near-equivalence of different AP–algorithms.

10.3. A Primal Method

In the following we introduce a primal algorithm for solving AP which has been developed by Balinski and Gomory [1964]. It can be viewed as an adaption of Klein's negative cycle method for general MCFP's since starting from an initial assignment successive improvements are made over negative alternating cycles. It is a non–simplex method since it maintains assignments i.e. feasible integer solutions but not basic solutions of the related LP.

Like the Hungarian method it was originally formulated for complete problems and all steps were performed on the cost–matrix. Balinski and Gomory showed the validity and finiteness of their procedure by using LP-arguments only and they claimed this method to be "dual to" the Hungarian method, while Gass [1975] called the procedure a "primal Hungarian method".

We will present this method here in a version which works on graphs and we emphasize on the combinatorial and graphical interpretation of the procedure.

Balinsky and Gomory's approach is based on the following theorem which is also valid for matchings in general non bipartite graphs and which in the case of bipartiteness can be viewed as a corollary of theorem 7.5. .

Theorem 10.5.

A perfect matching M in $G = (V, E)$ is optimal, i.e. is a min-cost perfect matching, iff G does not contain any negative M–alternating cycle.

Proof. Obviously the existence of a negative M–alternating cycle Q shows the non-optimality of M since

$$c(M \oplus Q) = c(M) + c(Q) < c(M).$$

Now assume the existence of a perfect matching M^* with $c(M^*) < c(M)$. Due to Theorem 9.1. every connected component of the graph $G' = (V, M \oplus M^*)$ is either an isolated node or a cycle the edges of which are alternately in M and M^*

143

(Note that G' cannot contain "alternating paths" since both matchings are perfect). If there is no such cycle then $M = M^*$, thus let $Q_1, \ldots, Q_p$ be the cycles in G' which induce M-alternating cycles in G with $M^* = M \oplus Q_1 \oplus \ldots \oplus Q_p$ and

$$c(M^*) = c(M) + c(Q_1) + \ldots + c(Q_p) < c(M).$$

Hence at least one of these cycles is negative with respect to M.

$\square$

This theorem motivates the following approach for finding a perfect matching (assignment) of minimal cost in a graph G, i.e. for solving 1MP resp. AP:

Procedure 10.1. **Negative cycle algorithm**

Step 0: Determine any perfect matching M.

Step 1: Find negative M–alternating cycle Q.

If no such cycle exists, STOP: M is a min–cost perfect matching.

Otherwise given such a cycle Q, goto Step 1.

Step 2: Set $M := M \oplus Q$ and goto Step 1.

If G is a bipartite graph negative M-alternating cycles can be determined in the following way:

The (undirected) graph $G = (V_s, V_t, E)$ is transformed into a directed graph $G' = (V_s, V_t, E')$ where every non-matching edge $\{i, j\} \in E \setminus M$ with $i \in V_s$ is directed from i to j and every matching edge $\{i, j\} \in M$ with $i \in V_s$ is directed from j to i. Then any directed cycle in G' corresponds to an M–alternating cycle in G and vice versa. Here the cost c_{ji} for edges directed from V_t to V_s are nonpositiv. Now negative M–alternating cycles in G resp. negative (directed)

144

cycles in G' can be found by applying label–correcting shortest path methods for instance. And this approach can be viewed as a possible implementation of Klein's method for AP.

Such a transformation resp. approach is not possible for nonbipartite problems. For that reason we present the labeling–method of Balinski and Gomory which works on the bipartite (undirected) graph G. This approach for finding negative alternating cycles can then be extended to the nonbipartite case, too (cf. section 13.3.).

From now on let $G = (V_s, V_t, E)$ be a bipartite graph and assume we are given an assignment M and two vectors $u, v \in \mathbb{R}^n$ fulfilling

$$u_i + v_j = c_{ij} \quad \text{for } \{i, j\} \in M.$$

We can interprete $(u, v) \in \mathbb{R}^{2n}$ as (possibly infeasible) solution for DAP which together with the assignment M fulfills the complementary slackness conditions. We call such a pair M and (u, v) *complementary* and we define the reduced cost of an edge as usual by $\bar{c}_{ij} := c_{ij} - u_i - v_j$.

Then the following lemma holds obviously.

Lemma 10.6.
Let M and (u, v) be a complementary pair and $Q = (V(Q), E(Q))$ be a M–alternating cycle. Then $c(Q) = \bar{c}(Q) = \sum_{(i,j) \in E(Q) \setminus M} \bar{c}_{ij}$.

Thus when checking for a negative M-alternating cycle we can as well use the reduced cost coefficients $\bar{c}_{ij}$.

Given a complementary pair M and (u, v) we define

$E(u, v) = \{\{i, j\} \in E \mid \bar{c}_{ij} = 0\}$ the set of admissible edges, and

$E^-(u, v) = \{\{i, j\} \in E \mid \bar{c}_{ij} < 0\}$ the set of negative edges.

Now the following observation is the basis for Balinski and Gomory's labeling method.

Let $j^* \in V_t$ such that $\delta(j^*) \cap E^-(u,v) \neq \emptyset$ and $i^* \in V_s$ such that $\{i^*, j^*\} \in M$. If no such node j^* exists, then we know that M is an optimal assignment. Otherwise choose edge $\{j^*, k^*\} \in \delta(j^*) \cap E^-(u,v)$. Now assume $T = (V(T), E(T))$ an M-alternating tree in $G(u,v) := (V, E(u,v))$ rooted at i^*. If $k^* \in O(T)$ then the alternating path P from i^* to k^* in T and the two edges $\{i^*, j^*\}$ and $\{j^*, k^*\}$ build an alternating cycle Q with respect to M with

$$c(Q) = \bar{c}(Q) = \bar{c}(P) + \bar{c}_{k^* j^*} - \bar{c}_{i^* j^*} = \bar{c}_{k^* j^*} < 0.$$

Hence Q is a negative M-alternating cycle.

In this case we perform the following "*mini-dual-change*":

$$\bar{v}_{j^*} := v_{j^*} - \bar{c}_{k^* j^*}$$

and we augment the matching M via the alternating cycle Q i.e.

$$\tilde{M} := M \oplus Q.$$

Then $\tilde{M}$ and $(\tilde{u}, \tilde{v})$ are a complementary pair again. The only effect of this dual change is that for each edge $e \in \delta(j^*)$ the reduced cost increases by the positive amount $(-\bar{c}_{j^* k^*})$.

Thus by this update no new negative edges are created, hence the total number of negative edges has decreased strictly.

Now assume that $T = (V(T), E(T))$ is a Hungarian tree in $G(u,v)$ rooted at i^* — i.e. T cannot be grown further in $G(u,v)$ — and $k^* \notin O(T)$. Then we determine (as in the Hungarian method)

$$\delta := min \ \{\bar{c}_{ij} \geq 0 \mid i \in O(T), j \notin V(T)\},$$

where we define $\delta = -\bar{c}_{k^* j^*}$ if the above set is void, and we perform the following *dual-change* steps similar to the Hungarian method:

146

$$\tilde{u}_i := \begin{cases} u_i + \delta, & \text{for } i \in O(T), \\ u_i, & \text{for } i \in V_s \setminus O(T), \end{cases}$$

$$\tilde{v}_j := \begin{cases} v_j - \delta, & \text{for } j \in I(T) \cup \{j^*\}, \\ v_j, & \text{for } j \in V_t \setminus (I(T) \cup \{j^*\}). \end{cases}$$

Then M and $(\tilde{u}, \tilde{v})$ form again a complementary pair, no new negative edges are created and all edges in $E(T)$ remain admissible. Moreover if $\delta \geq -\bar{c}_{k^* j^*}$ then the edge $\{k^*, j^*\}$ gets nonnegative reduced cost. Otherwise, if $\delta < -\bar{c}_{k^* j^*}$, the alternating tree T can be grown in $G(\tilde{u}, \tilde{v})$.

Thus after at most n dual changes either $k^* \in O(T)$ in which case edge $\{k^*, j^*\}$ gets zero reduced cost by a mini dual change and the matching is changed or $\{k^*, j^*\}$ gets nonnegative reduced cost while the matching remains unchanged. Hence the number of negative edges is decreased after at most n iterations.

Since at the beginning at most $n^2 - n$ negative edges exist an optimal assignment is obtained after $O(n^3)$ iterations. Every iteration can be performed in $O(n^2)$ steps — as in the Hungarian method — and we have a first rough estimation of $O(n^5)$ for the total complexity of this approach.

Now there are two principal possibilities of improving the above method:

- As for the Hungarian method improved updating formulas can be given to reduce the amount of work per iteration . We will discuss this possibility together with the respective improvements for the Hungarian method lateron.

- Moreover the number of iterations can be reduced significantly. We will shortly discuss two improvements in this direction which are of interest because of theoretical reasons, too.

So far the primal method did neither specify a "choice rule" for the next

147

negative edge to be chosen nor for the initial assignment. In the first refinement we will introduce a rule for choosing the next negative edge and the second refinement will additionally specify the starting solution and prescribe a modified labeling strategy.

Refinement 1 ("most-negative rule")

It is obvious that if edge $\{k^*, j^*\}$ is chosen in such a way that

$$\bar{c}_{k^* j^*} = min \; \{\bar{c}_{kj^*} \mid k \in N(j^*)\} < 0$$

then every edge in $\delta(j^*)$ has nonnegative reduced cost if the the edge (k^*, j^*) has become non–negative by a (mini–)dual change.

Thus choosing successively the "most-negative" edge incident with the nodes $j^* \in V_t$ gives an upper bound of n^2 iterations since at most n dual changes are necessary per node j^* to make all edges in $\delta(j^*)$ "nonnegative".

This refinement is close to the spirit of the "most negative" choice rules in simplex type methods.

Refinement 2 ("dimension expanding approach")

This second refinement which reduces the number of iterations to at most $n(n+1)/2$ is also of interest from a theoretical point of view. Let $i_{(1)}, \ldots, i_{(n)}$ resp. $j_{(1)}, \ldots, j_{(n)}$ be any ordering of elements in V_s resp. V_t with

$$c_{i_{(1)} j_{(1)}} = min \; \{c_{ij_{(1)}} \mid i \in V_s\}.$$

Then we start from the initial assignment

$$M^{(1)} = \{\{i_{(k)} j_{(k)}\} \mid k = 1, \ldots, n\}$$

and the dual solution

$$u_{i_{(k)}} = c_{i_{(k)} j_{(k)}} \qquad k = 1, \ldots, n$$
$$v_{j_{(k)}} = 0 \qquad k = 1, \ldots, n \; .$$

148

Hence $M^{(1)}$ and (u, v) form a complementary pair. Here we set $c_{i_{(k)}j_{(k)}} := M$ with M sufficiently large if $\{i_{(k)}, j_{(k)}\} \notin E$ and we keep those edges as "artifical" edges in E. Then we define the subgraphs $G^{(k)} := (V_s, V_t^{(k)}, E^{(k)})$ with

$$V_t^{(k)} := \{j_{(1)}, \dots, j_{(k)}\}$$
$$E^{(k)} := \{\{i, j\} \in E \mid j \in V_t^{(k)}\}.$$

At the beginning of our procedure all edges in $E^{(1)}$ have nonnegative reduced cost.

Let us assume that our calculation has reached a status with a complementary pair $M^{(k)}$ and $(u^{(k)}, v^{(k)})$ where all edges in $E^{(k)}$ have nonnegative reduced cost and $\{i_l, j_l\} \in M^{(k)}$ for $l > k$. Then two cases may occur:

<u>Case 1:</u> $min\ \{\bar{c}_{ij_{(k+1)}} \mid i \in V_s\} \geq 0$

In this case all edges in $E^{(k+1)}$ have nonnegative reduced cost.

<u>Case 2:</u> $\bar{c}_{i^* j_{(k+1)}} = min\ \{\bar{c}_{ij_{(k+1)}} \mid i \in V_s\} < 0$

In this case we choose edge $\{i^*, j_{(k+1)}\}$ to start the usual Balinski and Gomory labeling procedure. Yet we only grow the alternating trees in $G^{(k+1)}$ (resp. $G^{(k+1)}(u, v)$). Then after at most k iterations (dual–changes) all edges in $E^{(k+1)}$ have nonnegative reduced cost and a complementary pair $M^{(k+1)}$ and $(u^{(k+1)}, v^{(k+1)})$ is obtained with $\{i_l, j_l\} \in M^{(k+1)}$ for $l > k + 1$. Thus after at most $2 + 3 + \dots + n = n(n+1)/2$ iterations we end up with an optimal assignment in $G^{(n)} = G$.

We will analyse this second refinement further in section 10.6. when discussing the near equivalence of AP–approaches.

10.4. The AP–Simplex Method

In this section we introduce the *alternating basis (simplex-) method* for solving assignment problems which has been developed by Barr et al. [1977] and which is essentially a specialization of Cunningham's network simplex method (cf. section 7.2.).

The basic solutions of the linear program associated with AP correspond to spanning trees in the underlying graph and a basic solution assigns exactly n of the $2n - 1$ edges a value of 1 and the other $n - 1$ edges a value of zero. Thus the basic solutions of an AP are highly degenerate and this causes any simplex method to examine several bases (spanning trees) for the same solution (assignment) before moving to a "better" solution. As for MCFP such pivots are called *degenerate pivots*. In several computational studies it was observed that when solving large assignment problems by simplex methods, 95 percent of the pivots are degenerate (cf. Gavish and Schweitzer [1974], Barr et al. [1977]). Gavish et al. [1977] proposed special pivoting rules which reduce the number of degenerate pivots significantly. Yet the breakthrough came with the development of the network simplex method and its AP specialization the alternating basis simplex method.

Let T be a rooted basis tree for an assignment problem on the graph $G = (V_s, V_t, E)$ and let M be the associated assignment. Then T is called an *alternating path basis tree (AP-basis)* if

(i)   the root node r is a node from V_s ,

(ii)   the path from any node $v \in V_s \cup V_t$ to r in the tree is an M–alternating path.

Any assignment M in G can be "extended" to an AP-basis by adding exactly one non–matching edge to every node $v \in V_s \setminus \{r\}$ (eventually some artifical edges with sufficiently large cost ("Big M") have to be added).

Note that the above definition of an AP–basis which assumes that AP is given as a matching problem on an undirected graph is equivalent to the notion of strongly feasible trees when AP is formulated as flow problem on a directed graph.

Now let $T = (V, E(T))$ be an AP–basis and $e^* = \{i^*, j^*\} \notin E(T)$ with $i^* \in V_s$. Then "adding" e^* to T defines a unique cycle $C(T, e^*)$ and the following "leaving–edge rule" maintains the AP–basis property:

AP–leaving edge rule

Let $\bar{e} = \{i^*, j\}$ be the unique edge in $C(T, e^*)$ with $j \neq j^*$. Then $T' = (V, E(T'))$ with $E(T') = (E(T) \cup \{e^*\}) \setminus \{\bar{e}\}$ is again an alternating path basis tree.

With respect to an AP-basis tree T we can distinguish three types of non-tree edges. Here an edge $\{i^*, j^*\} \in E \setminus E(T)$ with $i^* \in V_s$ is called

- a *downward edge*, if i^* is a node of the alternating path from j^* to r,

- an *upward edge*, if j^* is a node of the alternating path from i^* to r,

- a *cross edge* otherwise.

Now it is easy to see that $e \in E \setminus E(T)$ gives rise to a nondegenerate pivot iff it is a downward edge.

Every downward edge $\{i^*, j^*\} \in E \setminus E(T)$ defines an M-alternating cycle Q which is composed from the alternating path $P_{i^* j^*}$ in T and edge $\{i^*, j^*\}$. Since all edges in $P_{i^* j^*}$ have reduced cost zero, $\bar{c}(Q) = \bar{c}_{i^* j^*}$ holds.

As already seen for the network simplex method only a subset of the dual variables change after a pivot and a simple updating rule can be formulated for the AP-simplex method, too.

Assume that the non-tree edge $\{i^*, j^*\}$ with $\bar{c}_{i^* j^*} < 0$ is to enter the basis and the edge $\{i^*, j\}$ is to leave the basis–tree. If $\{i^*, j\}$ is deleted from the tree T, two subtrees are formed. Let W denote the nodes of the subtree not containing r. Then the dual variables associated with the new basis–tree are obtained as follows:

$$\tilde{u}_i := \begin{cases} u_i - \delta, & \text{for } i \in V_s \cap W, \\ u_i, & \text{for } i \in V_s \setminus W, \end{cases}$$

$$\tilde{v}_j := \begin{cases} v_j + \delta, & \text{for } j \in V_t \cap W, \\ v_j, & \text{for } j \in V_t \setminus W, \end{cases}$$

$$\text{where } \delta := \begin{cases} -\bar{c}_{i^*,j^*}, & \text{if } i^* \in W, \\ \bar{c}_{i^*,j^*}, & \text{if } j^* \in W. \end{cases}$$

Now if $\{i^*, j^*\}$ induces a degenerate pivot then $i^* \in W$ and hence $\sum_{i \in V_s} u_i$ strictly decreases. Thus only a finite number of degenerate pivots may occur between two nondegenerate pivots, hence the AP-simplex method cannot cycle and the following theorem holds.

Theorem 10.7.

The AP–simplex method will obtain an optimal solution in a finite number of pivots regardless which "entering edge rule" is employed.

As for MCFP the strongly feasible network simplex method resp. AP–simplex method for assignment problems is not a "good" algorithm in the sense of Edmonds. Roohy–Loleh [1980] has given a class of AP's where the AP-simplex method performs an exponential number of combined degenerate and non–degenerate pivots.

Yet we have already mentioned in section 8.2. that transforming AP into a MCFP by adding an artifical source and an artifical sink and then solving the resulting problem by the strongly feasible network simplex method with "Big-M-start" and "most negative edge" entering rule leads to a polynomial algorithm for AP. In the following we will present two related entering rules which make the AP–simplex method a good algorithm independent of the choice of the starting basis–tree. The first result is due to Roohy-Loleh [1980].

<u>Refinement 1</u> ("altered most negative rule")

Let T^0 be an AP–basis tree and let M^0 and (u^0, v^0) be its associated assignment and dual solution, respectively. Let

$$E_+^0 := \{\{i,j\} \in E \mid c_{ij} - u_i^0 - v_j^0 \geq 0\}$$
$$E_-^0 := \{\{i,j\} \in E \mid c_{ij} - u_i^0 - v_j^0 < 0\}.$$

If $E_-^0 = \emptyset$, then T^0 resp. M^0 is an optimal (basic) solution. Otherwise enter the "most negative" edge $e_0 \in E_-^0$ into the basis to obtain a new AP–basis tree T^1 with the associated assignment M^1 and dual solution (u^1, v^1) . Now two cases may occur

<u>Case 1</u>: $c_{ij} - u_i^1 - v_j^1 \geq 0$ for all $e \in E_+^0$.

In this case we set $T^0 := T^1, M^0 := M^1$ and $(u^0, v^0) := (u^1, v^1)$ and we redefine E_+^0 and E_-^0 .

<u>Case 2</u>: Exists $\{i^*, j^*\} \in E_+^0$ with $c_{i^* j^*} - u_{i^*}^1 - v_{j^*}^1 < 0$.

In this case we perform simplex pivots by entering the most negative edges from the set E_+^0 until an AP-basis tree T^* and a dual solution (u^*, v^*) is obtained such that $c_{ij} - u_i^* - v_j^* \geq 0$ for all $\{i,j\} \in E_+^0$. Then we define $T^0 := T^*, M^0 := M^*$ and $(u^0, v^0) := (u^*, v^*)$ and we continue.

Let us define a *period* of the AP-simplex method with the altered most negative rule to be a sequence of pivots for which the set E_+^0 remains unchanged. Then the following theorem holds:

Theorem 10.8.   *(Roohy–Loleh [1980])*
The length of each period in the AP–simplex method with altered most negative rule is bounded from above by n .

During each period of the algorithm the cardinality of the set of negative edges is reduced by at least 1 and each period can have at most 1 nondegenerate pivot. At the beginning of the procedure at least all $2n - 1$ edges in the initial tree are nonnegative. Hence the following theorem holds.

Theorem 10.9. *(Roohy–Loleh [1980])*
In the AP–simplex method with altered most negative rule the total number of pivots is bounded by $n^3 - 2n^2 + n$ with $n^2 - 2n + 1$ the maximum possible number of degenerate pivots.

$\square$

Now the amount of work to perform one pivot–operation of the AP–simplex method with the altered most negative rule is of order $O(|E|)$ and thus the total complexity of the procedure becomes $O(n^3 |E|)$ or $O(n^5)$ which is inferior to the best known bounds of other AP algorithms.

Refinement 2 ("dimension expanding approach")

This refinement is similar to the refinement 2 of Balinski and Gomory's method introduced before.

Let $i_{(1)}, \ldots, i_{(n)}$ and $j_{(1)}, \ldots j_{(n)}$ be any ordering of the elements in V_s resp. V_t and $M^0 := \{\{i_{(k)}, j_{(k)}\} \mid k = 1, \ldots, n\}$ and define $E^{(q)} := \{\{i_{(k)}, j_{(l)}\} \mid 1 \le k, l \le q\}$. Let T^0 be an AP-basis tree corresponding to M^0 and (u^0, v^0) be its associated dual solution.

Then the AP-method is organized in n phases where at the beginning of phase q, $1 \le q \le n$, we define

$$E_+^0 := \{\{i, j\} \in E^{(q)} \mid c_{ij} - u_i^0 - v_j^0 \ge 0\},$$
$$E_-^0 := \{\{i, j\} \in E^{(q)} \mid c_{ij} - u_i^0 - v_j^0 < 0\}.$$

If $E_-^0 = \emptyset$, then the q th phase is completed. If $q = n$ then T^0 resp. M^0 is an optimal solution, else we start phase $q + 1$. If $E_-^0 \ne \emptyset$ we apply the AP-simplex method with altered most negative rule (refinement 1) using the above definition of the sets E_+^0 and E_-^0 to "solve" the $q + 1$ st phase subproblem.

The polynomiality of this second refinement follows immediately from the results of Roohy–Loleh for refinement 1.

While these two refinements of the AP–simplex method are polynomial in the problem size, Hung [1983] has given another variant of the AP–simplex method which is polynomial in the problem size and the encoding of the problem data.

Refinement 3 ("modified row most negative rule")

For this refinement assume that $i_1, \ldots, i_n$ and $j_1, \ldots, j_n$ are again arbitrary but fixed orderings of the sets V_s resp. V_t . In matrix notation for complete problems, the i's would correspond to the rows of the cost–matrix and the j's would correspond to columns, this motivates the name "row" most negative rule.

Let $\bar{e} = \{i, j\}$ be the entering edge that gives rise to the current AP–basis tree T . Now the modified row most negative rule would select $e^* = \{i+1, j^*\}$ as the next entering edge where $\bar{c}_{i+1,j^*} < 0$, e^* is an upward or cross edge with respect to T and $\bar{c}_{e^*}$ is minimal over all such edges. If every cross or upward nonbasic edge incident with node (row) $i+1$ has nonnegative reduced cost then one proceeds to "nodes" $i+2, i+3, \ldots, n, 1, 2, \ldots, i$ until such a pivot becomes possible. Note that if this rule succeeds to choose an edge e^* , the pivot will be degenerate. If this rule fails, i.e no degenerate pivot is possible, we call the current basis a *degenerate pivot free basis* and the entering edge e^* is determined by the most negative rule i.e.

$$\bar{c}_{e^*} = min \; \{\bar{c}_e \mid e \in E\} < 0.$$

The above refinement can also be described as follows: Given an AP–basis tree T with the associated assignment M, in a first phase degenerate pivots are performed (which only change the basis tree but not the assignment) until a degenerate-pivot-free basis–tree is obtained. Then either M is optimal or the next (nondegenerate) pivot must change the assignment, too.

From a theorem of Cunningham [1976] the following result follows immediately:

Lemma 10.10. *(Hung [1983])*
Any AP-basis tree T for an assignment M can be transformed into a degenerate pivot free AP–basis T' for M by at most $(n - 1)^2$ applications of the modified row most negative rule.

This property was combined by Hung with the following simple observation to show the polynomiality of this approach.

Lemma 10.11. *(Hung [1983])*
Let M^0 and M^* be, respectively, the current and the optimal assignment and $T_0(M^0)$ be any basis corresponding to M^0 with (u^0, v^0) the associated dual solution. Then

$$c(M^0) - c(M^*) \le \delta := -\sum_{i \in V_s} min\{c_{ij} - u_i^0 - v_j^0 \mid j \in V_t\}.$$

Then the following theorem shows the polynomiality of Hung's approach. Given a basis–tree T and its unique dual solution (u, v) we define $\bar{c}_{ij}(T) := c_{ij} - u_i - v_j$.

Theorem 10.12. *(Hung [1983])*
Let $T_0(M^0), \ldots, T_k(M^*)$ be the sequence of AP-bases generated by the AP–simplex method with modified row most negative rule, then $k \le n^3 \ln \Delta$ where $\Delta = c(M^0) - c(M^*)$.

156

Proof. Let T_m be the degenerate-pivot free basis for M^0 and let $\delta_0 = -\sum_{i \in V_s} min\{c_{ij}(T_m) \mid j \in V_t\}$. Since the entering edge e^* is chosen to have the most negative reduced cost it follows that $\bar{c}_{e^*}(T_m) \leq -(1/(n-1))\delta_0$. Thus with M^1 the matching obtained after the nondegenerate pivot using e^* we get

$$c(M^1) - c(M^*) = c(M^0) - c(M^*) + \bar{c}_{e^*}(T_m)$$
$$\leq \Delta - (1/(n-1))\delta_0 \leq ((n-2)/(n-1)) \cdot \Delta.$$

Applying this argument inductively to the sequence $M^0, M^1, M^2, \ldots, M^t = M^*$ of matchings generated by the procedure yields

$$c(M^t) - c(M^*) \leq ((n-2)/(n-1))^t \cdot \Delta$$

where $M^t = M^*$ if $c(M^t) - c(M^*) < 1$. Thus t is the least integer satisfying $t > log\,\Delta / log((n-1)/(n-2))$ for any base of the logarithm. Now one can show that $ln((n-1)/(n-2)) > 1/(n-1)$ holds. Since at most $(n-1)^2$ degenerate pivots for a fixed assignment M^i are possible, we get for $ln\,\Delta \geq 1/2$ (implying $\Delta \geq 2$)

$$k \leq (n-1)^2 \cdot (1 + (n-1)\,ln\,\Delta) < n^3\,ln\,\Delta.$$

$\square$

10.5. The Shortest Augmenting Path Method

In this section we introduce the shortest augmenting path (SAP) approach for solving assignment problems. Several methods proposed in literature fall immediately under this concept. Early references are Dinic and Kronrod [1969] as well as Tomizawa [1972] and Edmonds and Karp [1972] who introduced this concept for solving general network flow problems. A first ALGOL–program of Tomizawa's method was given by Dorhout [1973].

Recently the SAP–approach was (re–)discovered by several authors (cf. Hung and Rom [1980] and Enquist [1983]).

In this section we will show that all these methods are only refinements of a basic SAP–method and not entirely different methods as some of the authors claim.

In fact we will show in the next section that all the common AP-algorithms can be viewed as special SAP–implementations. Yet the relationship of the above mentioned algorithms is even closer.

In section 7.4. we have defined an s-t-flow x to be extreme if it is of minimum cost among all flows with the same value $z(x)$. Lemma 7.10. stated that this is equivalent to the property that $G(x)$ contains no negative directed cycle. Flows having this property can also be thought of as "local optimal" if the neighbourhood–topology is defined properly.

For matching problems we will use the same concept and we define a matching M in a (not necessarily bipartite) graph $G = (V, E)$ to be *extreme* iff M does not allow any negative M–alternating cycle Q , i.e. a M–alternating cycle $Q = (V(Q), E(Q))$ with

$$c(E(Q) \setminus M) < c(E(Q) \cap M) \quad \text{resp.}$$

$$c(M \oplus Q) < c(M).$$

Note that here an extreme matching need not be of minimal cost among all matchings having the same cardinality.

Hence we distinguish several special classes of extreme matchings:

A matching M in G is called *k-optimal* iff $|M| = k$ and $c(M) \leq c(M')$ for all matchings M' in G with $|M'| = k$.

If $G = (V_s, V_t, E)$ is bipartite we call a matching M to be $V_s(M)$-optimal ($V_t(M)$-optimal) iff $c(M) \leq c(M')$ for all matchings M' in G with $V_s(M') = V_s(M)$ ($V_t(M') = V_t(M)$) .

The shortest augmenting path method for solving AP's is based on the following theorem which is valid for general (not necessarily bipartite) graphs.

Theorem 10.13.

Let M be an extreme matching in $G = (V, E)$ and let P be the shortest M-augmenting path connecting two M-unsaturated nodes s and t .

Then $M' := M \oplus P = (M \setminus E(P)) \cup (E(P) \setminus M)$ is an extreme matching in G .

Proof.   Assume M' is not extreme, then exists a negative M'–alternating cycle $Q = (V(Q), E(Q))$. Now $E(Q) \cap E(P) \neq \emptyset$ since otherwise Q would also be a negative M–alternating cycle. Now let $\tilde{E} := E(P) \oplus E(Q)$ $(= (E(P) \setminus E(Q)) \cup (E(Q) \setminus E(P)))$. Then $\tilde{E}$ induces a M-augmenting path P' with

$$M \oplus P' = (M \oplus P) \oplus Q .$$

Hence $c(M \oplus P') < c(M \oplus P)$ from which $c(P') < c(P)$ follows, which is a contradiction to the assumption that P is a shortest M–augmenting path.

$\square$

In the bipartite case the following Corollary follows immediately.

Corollary 10.14.

Let M be a $V_s(M)$-optimal matching in a bipartite graph $G = (V_s, V_t, E)$ and let $s^* \in V_s \setminus V_s(M)$. Let P be the shortest M–augmenting path connecting

s^* with a node $j \in V_t \setminus V_t(M)$. Then $M' = M \oplus P$ is $V_s(M')$–optimal with $V_s(M') = V_s(M) \cup \{s\}$.

The following theorem holds in general (not necessarily bipartite) graphs again.

Theorem 10.15.

Let M be a k–optimal matching in $G = (V, E)$ and let P be the shortest M–augmenting path. Then $M' = M \oplus P$ is $(k+1)$–optimal.

Proof. Assume any $(k+1)$–optimal matching M^*, then $M^* = M \oplus K_1 \oplus \ldots \oplus K_p$ with K_i, $i = 1, \ldots, p$, cycles and paths the edges of which are alternating with respect to M and M^*. For any cycle and even path K_i we have $c(K_i \cap M) = c(K_i \cap M^*)$ since otherwise M would not be k–optimal resp. M^* would not be $(k+1)$–optimal. Thus w.l.o.g. we may assume that all the K_i's are odd paths. The number of M–augmenting paths among the K_i's must be exactly one larger than the number of M^*–augmenting paths. Now assume K_{i_1} a M^*–augmenting path and K_{i_2} a M–augmenting path then

$$c(M \oplus K_{i_1} \oplus K_{i_2}) = c(M^* \oplus K_{i_1} \oplus K_{i_2})$$

since otherwise M would not be k–optimal resp. M^* would not be $(k+1)$–optimal. Thus $c(M^*) = c(M \oplus K_1 \oplus \ldots \oplus K_p) \geq c(M \oplus P)$.

$\square$

Obviously reverse statements hold, too:

Lemma 10.16.

Let M be an extreme matching and P an M–augmenting path connecting s and t.

(i) If $M \oplus P$ is extreme, then P is the shortest M-augmenting path connecting s and t.

(ii)   If M is k–optimal and $M \oplus P$ is $(k+1)$– optimal then P is the shortest M–augmenting path.

(iii)   If $G = (V_s, V_t, E)$ is a bipartite graph with $s \in V_s$ and $M \oplus P$ is $V_s(M \oplus P)$-optimal then P is the shortest M-augmenting path connecting s to a node in $V_t \setminus V_t(M)$.

Proof.   (ii) and (iii) hold obviously and we only have to show (i). Assume exists a M–augmenting path P' connecting s and t with $c(P') < c(P)$. Then $Q = (P' \oplus P)$ is a $(M \oplus P)$–alternating cycle with

$$(M \oplus P) \oplus Q = M \oplus P'.$$

Thus $c(Q) < 0$ which is a contradiction to the assumption that $M \oplus P$ is extreme.

□

Now assume an extreme matching M in a bipartite graph $G = (V_s, V_t, E)$. Then M is obviously an optimal solution to the linear program

$$min \sum_{\{i,j\} \in E} c_{ij} \cdot x_{ij} \qquad \text{subject to}$$

$$\sum_{j \in V_t} x_{ij} = \begin{cases} 1, & \text{for } i \in V_s(M) \\ 0, & \text{else} \end{cases}$$

$$\sum_{i \in V_s} x_{ij} = \begin{cases} 1, & \text{for } j \in V_t(M) \\ 0, & \text{else} \end{cases}$$

$$x_{ij} \in \{0,1\} \quad \text{for } \{i,j\} \in E.$$

Thus exists a dual solution (u, v) such that

$$u_i + v_j \le c_{ij} \quad \text{for all } \{i,j\} \in E,$$

$$u_i + v_j = c_{ij} \quad \text{for } \{i,j\} \in M.$$

As for extreme flows we call such a pair M and (u,v) *compatible* and we consider the reduced costs $\bar{c}_{ij} = c_{ij} - u_i - v_j$ for $\{i,j\} \in E$. For P an M–augmenting path connecting $i_0 \in V_s \setminus V_s(M)$ and $j_0 \in V_t \setminus V_t(M)$ the following relation holds

$$\bar{c}(P) = \Big(\sum_{\{i,j\} \in E(P) \setminus M} c_{ij} - u_i - v_j \Big) - \Big(\sum_{\{i,j\} \in E(P) \cap M} c_{ij} - u_i - v_j \Big)$$

$$= c(P) - u_{i_0} + v_{j_0}.$$

Thus as for the negative cycle approach we can use the reduced cost coefficients instead of the original cost coefficients when searching for the shortest augmenting path connecting two prespecified unsaturated nodes. But with respect to the reduced costs the cost definiton for alternating paths becomes a (simple) sum of nonnegative weights:

$$\bar{c}(P) = \sum_{\{i,j\} \in E(P)} \bar{c}_{ij}.$$

Thus the shortest augmenting path can be found by efficient label-setting techniques, like the Dijkstra-procedure for instance. Note that only from this point on the notation "shortest" augmenting path makes sense, while before we should have better used the term "least (marginal) cost" augmenting path.

Moreover it can be shown that for $V_s(M)$–optimal matchings M a dual solution exists which is compatible with M and fulfills

$$u_i = 0 \qquad \text{for } i \notin V_s(M),$$

$$v_j = 0 \qquad \text{for } j \notin V_t(M).$$

In this case we call M and (u,v) a $V_s(M)$-*compatible pair*. In the following we will formulate a basic labeling method which finds shortest–augmenting paths with respect to an extreme matching M if a compatible dual solution is at hand. Thereafter we will discuss the use of this basic procedure within several variants of the shortest augmenting path approach.

This method grows a M–alternating tree from a prespecified node $i_0 \in V_s \setminus V_s(M)$ such that for each node $k \in V$ contained in the tree the alternating path to the root i_0 is the shortest M–alternating path connecting i_0 and k.

As in the CMP–labeling method, nodes which are contained in the alternating tree are (permanently) labeled while nodes not in the tree are unlabeled or tentatively labeled.

In fact we use two different types of labels. Any node $k \in V$ is assigned a d–label which gives the length of the shortest alternating path from i_0 to k (with respect to the reduced cost) found so far and each $j \in V_t$ with finite d-label is assigned a label $p(j) \in V_s$ which points to the predecessor on this alternating path from j to i_0 .

In the course of the procedure we keep a list L of "candidate" nodes in V_t , i.e. nodes with finite d–value. Then we will permanently label the node j^* in L with minimal d–value, like in Dijkstra's shortest path method. If $j^* \in V_t \setminus V_t(M)$, we have found a shortest augmenting path connecting i_0 and j^* . Otherwise we will introduce j^* and the node $i^* \in V_s$ which is matched with j^* into the tree and use the d-value of j^* to update L and the d-values for non-tree nodes in V_t .

Procedure 10.2.   **SAP–labeling procedure (alternating tree version)**

Step 0:   Set

$$d_i := \begin{cases} 0, & \text{for} \quad i = i_0 \\ \infty, & \text{for} \quad i \in V_s \setminus \{i_0\} \end{cases}$$

$$d_j := \begin{cases} \bar{c}_{i_0,j}, & \text{for} \quad j \in N(i_0) \\ \infty, & \text{for} \quad j \in V_t \setminus N(i_0) \end{cases}$$

$$p_j := i_0 \quad \text{for } j \in N(i_0)$$

$$L := N(i_0).$$

Step 1: If $L = \emptyset$, STOP: no M–augmenting path exists.

Otherwise determine $j^* \in L$ with $d^* := d_{j^*} = min\{d_j \mid j \in L\}$.

If $j^* \in V_t \setminus V(M)$, STOP: a shortest M-augmenting path P connecting i_0 and j^* has been found.

Otherwise goto Step 2.

Step 2: Let $i^* \in V_s$ with $\{i^*, j^*\} \in M$.

Set $L := L \setminus \{j^*\}$ and $d_{i^*} := d^*$.

(i^* and j^* become permanently labeled)

For $j \in N(i^*)$, not permanently labeled, determine

$d_j := min\{d_j, \bar{c}_{i^* j} + d^*\}$

and set $p(j) := i^*$ if $d_j = d^* + \bar{c}_{i^* j}$.

Goto Step 1.

If the method stops in Step 1 we know that AP has no feasible solution. Otherwise it halts with a (shortest) augmenting path P . In that case the d–values of permanently labeled nodes give the length of shortest alternating paths starting from $i_0 \in V_s$. After the augmentation these d–values can be used to construct a dual solution $(\tilde{u}, \tilde{v})$ which together with $\bar{M} := M \oplus P$ forms a compatible pair again.

For that purpose define

$$\tilde{u}_i := u_i + (d^* - d_i) \quad \text{for } i \in V_s \quad \text{permanently labeled,}$$

$$\tilde{v}_j := v_j - (d^* - d_j) \quad \text{for } j \in V_t \quad \text{permanently labeled.}$$

It is an easy exercise to show that $\bar{M} := M \oplus P$ and $(\tilde{u}, \tilde{v})$ are compatible. Thus if M is not perfect another iteration can be started using the associated reduced costs. The complexity of the above labeling procedure and the consecutive dual update is $O(|E|)$. Since at most $|V_s|$ augmentations are necessary we get a total complexity of $O(|V_s| \cdot |E|)$. In computer implementations the nodes in L may be stored in a *priority queue* ordered by their d–values. By priority queue we mean any list-structure which allows insertion and deletion of elements to be performed in $O(\log n)$ steps where n is the length of the list (cf. Aho et. al [1974]).

An immediate modification of this labeling procedure is the socalled alternating–forest variant where we would start to grow alternating trees from all nodes $i \in V_s \setminus V_s(M)$ simultaneously. Then only Step 0 has to be modified:

Step 0': Set

$$d_i := \begin{cases} 0, & \text{for} \quad i \in V_s \setminus V_s(M) \\ \infty, & \text{for} \quad i \in V_s(M) \end{cases}$$

$$L := \bigcup_{i \in V_s \setminus V_s(M)} N(i)$$

$$d_j := \begin{cases} \min\{\bar{c}_{ij} \mid \{i,j\} \in E, i \in V_s \setminus V_s(M)\} & \text{for } j \in L \\ \infty & \text{else} \end{cases}$$

$$p(j) := i_0 \text{ if } \bar{c}_{i_0 j} = d_j \text{ for } j \in L$$

Now it is easy to see that starting with a $V_s(M)$–optimal pair both variants — the tree–variant and the forest–variant — will produce $V_s(M \oplus P)$–compatible pairs with eventually different sets $V_s(M \oplus P)$ due to different shortest augmenting paths which have been found.

We now describe two special variants which are of special theoretical interest:

165

<u>Refinement 1</u>   ("dimension expanding variant")

Let $i_{(1)}, \ldots i_{(n)}$ be an ordering of V_s and $j_{(1)}, \ldots, j_{(n)}$ be an ordering of V_t. We assume that $\{i_{(k)}, j_{(k)}\} \in E$ for $k = 1, \ldots, n$ where we add artificial edges with sufficiently large cost if necessary. The subgraphs $G^{(k)} = (V_s^{(k)}, V_t^{(k)}, E^{(k)})$ for $1 \le k \le n$ are defined as follows:

$$V_s^{(k)} := \{i_{(1)}, \ldots, i_{(k)}\}$$
$$V_t^{(k)} := \{j_{(1)}, \ldots, j_{(k)}\}$$
$$E^{(k)} := \{\{i, j\} \in E \mid i \in V_s^{(k)}, j \in V_t^{(k)}\}$$

Now let $M^{(k)}$ be an optimal assignment in $G^{(k)}$. Then $M^{(k)}$ is an extreme matching in $G^{(k+1)}$. Hence an optimal assignment $M^{(k+1)}$ in $G^{(k+1)}$ can be obtained from $M^{(k)}$ by augmenting along the shortest augmenting path starting at $i_{(k+1)} \in V_s^{(k+1)}$. Now let $(u, v) \in \mathbb{R}^{2k}$ be the dual solution which is compatible with $M^{(k)}$. Setting

$$u_{i_{(k+1)}} := min\{min\{c_{i_{(k+1)},j} - v_j \mid j \in V_t^{(k)}\}, c_{i_{(k+1)},j_{(k+1)}}\}$$
$$v_{j_{(k+1)}} := min\{min\{c_{i,j_{(k+1)}} - u_i \mid i \in V_s^{(k)}\}, 0\}$$

yields an $M^{(k)}$-compatible dual solution in $G^{(k+1)}$. Thus the optimal matching $M^{(k+1)}$ can be found by one application of the SAP–tree–version.

<u>Refinement 2</u>   ("k–optimal version")

A sequence of k-optimal extreme matchings $M^{(1)}, M^{(2)}, \ldots, M^{(n)}$ is constructed by the forest–version of the SAP–labeling approach when starting from $M^{(0)} = \emptyset$ and $u_i = v_j = 0$ for $i \in V_s, j \in V_t$.

This can easily be seen by the following arguments.
Let $M^{(k)}$ be the extreme matching obtained after k iterations and $(u^{(k)}, v^{(k)})$ the compatible dual solution. Now all nodes in $V_s \setminus V_s(M^{(k)})$ are labeled in each iteration hence $u_i^{(k)} = u_j^{(k)} = \Delta$ for $i, j \in V_s \setminus V_s(M^{(k)})$. Analogously we have $v_j^{(k)} = 0$ for all $j \in V_t \setminus V_t(M^{(k)})$, since these nodes were never labeled.

The problem of finding a k–optimal assignment can be formulated as follows

$$min \ \{c'x \mid Ax \leq 1, 1'x = k, x \geq 0\}$$

with A the node–edge incidence matrix of G .

Then the associated dual program is

$$max \ \sum_{i \in V_s} u_i + \sum_{j \in V_t} v_j + \lambda$$

$$u_i + v_j + \lambda \leq c_{ij} \quad \text{for } \{i, j\} \in E$$

$$u_i, v_j \leq 0 \quad \text{for } i \in V_s, j \in V_t \ .$$

From complementary slackness we know that $M^{(k)}$ is k–optimal iff exists a dual solution $(u*, v*, \lambda*) \in \mathbb{R}^{2n+1}$ with the property

$$i \in V_s \backslash V_s(M*) \Rightarrow u_i^* = 0$$

$$j \in V_t \backslash V_t(M*) \Rightarrow v_j^* = 0$$

$$\{i, j\} \in M^* \Rightarrow u_i^* + v_j^* + \lambda = c_{ij}.$$

Now setting

$$u_i^* := u_i^{(k)} - \Delta \quad \text{for } i \in V_s$$

$$v_j^* := v_j \quad \text{for } j \in V_t$$

$$\lambda^* := \Delta$$

yields the desired relations.

$\square$

10.6. Near Equivalence of AP–Algorithms

From the results on network flow algorithms we know already that all the AP–methods introduced before can be viewed as only different implementations of the same basic principle — the shortest augmenting path approach. In the following we want to outline only some interesting properties and relations for the AP–methods.

Another look at the Hungarian method shows that all the matchings produced in the course of the procedure are extreme. This is easily seen since a compatible dual solution is always constructed. More precisely: the dual solution is given first and then a compatible matching of maximal cardinality is constructed. Moreover if we start the primal–dual approach with $u_i = v_j = 0$ for $i \in V_s, j \in V_t$, then a sequence of k–optimal solutions will be constructed.

Let us now analyse the labeling process of the Hungarian method by looking at the sequence of steps during one phase, i.e. between two augmentations. For that purpose let $M^{(k)}$ and $(u^{(k)}, v^{(k)})$ be the extreme matching and compatible dual solution, respectively, at the beginning of a phase. By $\bar{c}_{ij}^{(k)}$ we denote the reduced cost of an edge $\{i, j\}$ at the beginning of the phase. After another r dual-updates, with δ–values $\delta_1^*, \ldots, \delta_r^*$ say, the dual variables are changed to

$$u_i := u_i^{(k)} + \Delta_i^{(r)} \quad \text{for } i \in V_s$$

$$v_j := v_j^{(k)} - \Delta_j^{(r)} \quad \text{for } j \in V_t$$

For each node $q \in V$ we denote by $L(q)$ the number of dual updates during the present phase which were performed <u>before</u> node q became labeled, i.e. a member of the alternating forest, and $L(q) := r$ if q is unlabeled. Now define

$$D := \sum_{l=1}^{r} \delta_l^* \quad \text{and}$$

$$d_q := \sum_{l=1}^{L(q)} \delta_l^*$$

then the following relation holds

$$\Delta_q^{(r)} = D - d_q \quad \text{for } q \in V.$$

In the next step the Hungarian method would determine the δ-value δ_{r+1}^* according to the following formula

$$\delta_{r+1}^* = \min \{\bar{c}_{ij} \mid i \text{ labeled }, j \text{ unlabeled }\}$$
$$= \min \{c_{ij} - (u_i^{(k)} + \Delta_i^{(r)}) - (v_j^{(k)} - \Delta_j^{(r)}) \mid i \text{ labeled }, j \text{ unlabeled }\}.$$

Since $\Delta_j^{(r)} = 0$ for j unlabeled this is equivalent to

$$\delta_{r+1}^* = \min \{c_{ij} - u_i^{(k)} - v_j^{(k)} - (D - d_i) \mid i \text{ labeled }, j \text{ unlabeled }\}$$
$$= \min \{c_{ij} - u_i^{(k)} - v_j^{(k)} + d_i \mid i \text{ labeled }, j \text{ unlabeled }\} - D.$$

Now instead of determining δ_{r+1}^* we can equivalently determine

$$\hat{\delta}_{r+1}^* = \min \{d_i + \bar{c}_{ij}^{(k)} \mid i \text{ labeled }, j \text{ unlabeled }\}.$$

But this formula is exactly the formula for determining the minimum in the SAP–labeling method. Hence the SAP–labeling technique can also be viewed as a more economical implementation of the Hungarian method where several dual updates are comprised into one "general" update at the end of a phase after the augmentation has been performed.

After showing the (even more than) near–equivalence between the Hungarian method and the SAP–method we want to discuss the relations between the SAP–approach and the refinements of Balinski and Gomory's primal method.

In refinement 1 ("most negative edge rule") we successively choose nodes $j^* \in V_t$ and $k^* \in V_s$ such that $\bar{c}_{k^* j^*} = \min \{\bar{c}_{ij^*} \mid i \in N(j^*)\} < 0$. After at most n dual changes all edges in $\delta(j^*)$ get nonnegative reduced cost. Thus after k iterations let $V_t^{(k)} \subseteq V_t$ be the set of nodes j^* which have been chosen in the first k iterations and let M be the current assignment. Then $M^{(k)} := \{\{i,j\} \in M \mid j \in V_t^{(k)}\}$ is an extreme matching, more precisely a $V_t^{(k)}$–optimal matching. Hence if we choose $j^* \in V_t \setminus V_t^{(k)}$ for the next iteration, the primal method will produce a $V_t^{(k+1)}$-optimal matching. Since $V_t^{(k+1)} = V_t^{(k)} \cup \{j^*\}$ the method implicitly augments $M^{(k)}$ along the shortest $M^{(k)}$–augmenting path starting at j^* . This shortest augmenting path may either be the single edge $\{i^*, j^*\} \in M$ in which case no "real" augmentation is performed. Otherwise the shortest $M^{(k)}$–augmenting path P starting at j^* is given by the $V(P) = V(Q)$ and $E(P) = E(Q) \setminus \{i^*, j^*\}$ where Q is the negative M–alternating cycle found by the primal method.

In a similar way one can easily show that refinement 2 of Balinski and Gomory's approach and the dimension expanding variant of the SAP–approach are equivalent.

Finally we want to analyse the near equivalence of assignment–simplex methods and the shortest augmenting path method. We have already mentioned in section 8.2. that applying the network-simplex method with "Big-M-start" and "most negative entering rule" to the MCFP obtained from AP by adding an artifical source and sink is equivalent to the shortest augmenting path method. In fact by this approach successively k–optimal matchings in the original graph are constructed. Roohy–Loleh [1980] has shown that this method is polynomial if strongly feasible trees are maintained. Let us focus now on refinement 2 of the AP–simplex method. Here we denote by $M^{(q)}$ the assignment obtained at the end of the q-th phase. Then it is evident that the matching $\tilde{M}^{(q)} = M^{(q)} \cap E^{(q)}$ is an optimal assignment in the subgraph induced by the "first" q nodes in V_s and V_t , respectively. Thus this variant of the AP-simplex method is (near–)equivalent to the dimension expanding SAP–approach.

After showing the near equivalence of the four basic AP–approaches we want to mention shortly some results on the practical efficiency of these methods.

An early computational study of Florian and Klein [1970] revealed that Balinski and Gomory's approach is inferior to the Hungarian method. First comparisons between the Hungarian method and Tomizawa's SAP implementation were performed by Dorhout [1973] showing the SAP–concept to be superior. An alternating–tree version of the SAP–approach was published by Burkard and Derigs [1980]. This implementation was shown to be significantly superior to Dorhout's code.

Thompson [1982] compared the Dorhout-code and the Burkard–Derigs–code with an implementation of his recursive method (cf. Thompson [1981]) which can also be viewed as a special implementation of the SAP-approach. Carpaneto and Toth [1980], [1983] published computer codes of the Hungarian method which they showed to be superior to other published codes of the Hungarian method. An anonymous referee of Carpaneto and Toth's 1980–paper contributed results from test series showing Carpaneto and Toth's method to be inferior to the alternating basis simplex code of Barr et al. [1977]. Hung and Rom [1980] developed a special implementation of the SAP–approach which they call the "relaxation method" and they claimed their method to be twice as fast as the AP–simplex method. Enquist [1982] reinvented the SAP–approach and he compared his code with the code of Barr et al., Hung and Rom and another SAP–implementation by Weintraub and Barahona [1979]. The results showed his SAP–approaches to outperform the other implementations. In a recent paper McGinnis [1983], incorporating ideas of Bertsekas [1981], showed that clever implementations of the Hungarian method outperform simplex type methods. At last Glover et al. [1983] presented the "threshold assignment algorithm" a new SAP-implementation which uses a special label correcting shortest path algorithm. Results are reported which indicate that this code is significantly faster than Enquist's code for instance. In a recent study we could show that an improved version of our SAP–implementations is outperforming the codes of Thompson [1981] and Carpaneto and Toth [1980], [1981] as well as the primal approach and primal–dual code of McGinnis [1983]. A comparison with Enquist's code and the threshold assignment code was not possible since these codes are not published and not available.

Chapter 11. The Hitchcock Transportation Problem

In this section we discuss the second key problem within the class of general matching problems — the Hitchcock transportation problem (HTP). We have already shown that this problem can be viewed either as the specialization of the min–cost–flow problem on a directed bipartite graph or the specialization of the b–matching problem on a (nondirected) bipartite graph.

Here we treat HTP as a special matching problem on a bipartite graph $G = (V_s, V_t, E)$. Then HTP can be formulated as a linear program in the following way:

$$
min \sum_{\{i,j\} \in E} c_{ij} x_{ij}
$$

$$
\sum_{j \in N(i)} x_{ij} = b_i \quad \text{for } i \in V_s
$$

$$
\sum_{i \in N(j)} x_{ij} = b_j \quad \text{for } j \in V_t
$$

$$
x_{ij} \geq 0 \quad \text{for } \{i,j\} \in E.
$$

The set of feasible solutions is denoted by $P(A, b)$ and called the *transportation polytope* associated with A and b . From the theorem of Hoffmann and Kruskal [1956] we know that $P(A, b)$ has integer valued extreme points if b is integer valued. Various aspects of the transportation polytope are discussed by Klee and Witzgall [1968].

In section 5.3. we have already shown that any HTP on a bipartite graph $G = (V_s, V_t, E)$ with integer capacities $b_i, i \in V_s$, and $b_j, j \in V_t$, can be transformed into an AP on a related graph $G' = (V_s', V_t', E')$ where for every node $i \in V_s$ ($j \in V_t$) we have introduced b_i resp. b_j copies and two "copies" i' and j' are connected in G' by an edge iff $\{i, j\} \in E$. Thus HTP can be solved by applying any of the AP–algorithm to the assignment problem on G' .

Yet all AP–approaches have been extended to work on G directly and one can observe that the HTP–algorithms simply comprise into one step several successive steps the AP–versions would have performed on G'. From this observation and the results in section 10.6. it follows that the common HTP algorithms can again be interpreted as only different implementations of the shortest augmenting path principle. In a sense one can interpret the HTP–steps as "multi–augmentations" only.

In the following we will therefore only outline the shortest augmenting path approach for HTP and then show how this approach combined with an anticipant augmentation strategy — the so–called scaling technique — leads to a polynomial algorithm for HTP (and hence to a polynomial algorithm for MCFP, too).

11.1. The Shortest Augmenting Path Approach

For the following discussion we call a vector $x : E \rightarrow \mathbb{N} \cup \{0\}$ a *transportation vector* if

$$\sum_{j \in N(i)} x_{ij} \leq b_i \quad \text{for } i \in V_s$$

$$\sum_{i \in N(j)} x_{ij} \leq b_j \quad \text{for } j \in V_t$$

$$x_{ij} \geq 0 \quad \text{for } \{i, j\} \in E.$$

With respect to a transportation vector x we define

$$E^0(x) := \{\{i, j\} \in E \mid x_{ij} = 0\}$$

$$E^+(x) := E \setminus E^0(x)$$

$$b_i(x) := \sum_{j \in N(i)} x_{ij} \quad \text{for } i \in V_s$$

$$b_j(x) := \sum_{i \in N(j)} x_{ij} \quad \text{for } j \in V_t$$

$$b(x) := \sum_{\{i,j\} \in E} x_{ij} \quad \text{the ``value'' of } x$$

$$c(x) := \sum_{\{i,j\} \in E} c_{ij} \cdot x_{ij} \quad \text{the ``cost'' of } x$$

$$U_s(x) := \{i \in V_s \mid b_i(x) < b_i\} \quad \text{the set of } x\text{--unsaturated nodes in } V_s$$

$$U_t(x) := \{j \in V_t \mid b_j(x) < b_j\} \quad \text{the set of } x\text{--unsaturated nodes in } V_t$$

Now a path (or cycle) $P = (V(P), E(P))$ is called x–alternating if exists a bipartition of $E(P)$ into $E^+(P) \cup E^-(P)$ such that P is alternating with respect to this partition and $E^-(P) \subseteq E^+(x)$. An x–alternating path P connecting $i_0 \in U_s(x)$ and $j_0 \in U_t(x)$ is called x–augmenting if $\delta(i_0) \cap E^-(P) = \delta(j_0) \cap E^-(P) = \emptyset$.

For an x–alternating cycle $P = (V(P), E(P))$ we define

$$\kappa(P) = min\ \{x_{ij} \mid \{i,j\} \in E^-(P)\}$$

and for an x–augmenting path $P = (V(P), E(P))$ connecting $i_0 \in U_s(x)$ and $j_0 \in U_t(x)$ we define

$$\kappa(P) = min\{min\{x_{ij} \mid \{i,j\} \in E^-(P)\}, b_{i_0} - b_{i_0}(x), b_{j_0} - b_{j_0}(x)\},$$

then for all $\delta \in \{1, 2, \ldots, \kappa(P)\}$ the vector $y : E \rightarrow \mathbb{N} \cup \{0\}$ defined by

$$y_{ij} := \begin{cases} x_{ij} + \delta, & \text{for } \{i,j\} \in E^+(P) \\ x_{ij} - \delta, & \text{for } \{i,j\} \in E^-(P) \\ x_{ij}, & \text{otherwise} \end{cases}$$

is again a transportation vector. The above transformation is called the augmentation of x via P and will be denoted by

$$y := x \bigoplus \delta \cdot x(P).$$

With $c(P) := c(E^+(P)) - c(E^-(P))$ the (incremental) cost of P we get

$$c(y) = c(x) + \delta \cdot c(P) \qquad \text{and}$$

174

$$b(y) = \begin{cases} b(x), & \text{if } P \text{ is an x–alternating cycle} \\ b(x) + \delta, & \text{if } P \text{ is an x–alternating path.} \end{cases}$$

In analogy to the AP–case we call a transportation vector *extreme* if no x–alternating cycle Q with $c(Q) < 0$, a so–called *negative alternating cycle*, exists. Again we can distinguish different special types of extreme transportation vectors. A transportation vector x is called *k–optimal* if $b(x) = k$ and $c(x) \leq c(y)$ for all transportation vectors y with $b(y) = k$.

A transportation vector is called $b_s(x)$-*optimal* if $c(x) \leq c(y)$ for all transportation vectors y with $b_i(y) = b_i(x)$ for $i \in V_s$.

Now the following theorem which is essentially a corollary of the corresponding theorems for AP is the basis for the shortest alternating path approach for HTP:

Theorem 11.1.

(i) Let x be an extreme transportation vector, $i_0 \in U_s(x)$ and $j_0 \in U_t(x)$ and let P be a shortest x–augmenting path connecting i_0 and j_0 . Then for any $\delta \in \{1, \ldots, \kappa(P)\}$ the transportation vector

$$y := x \oplus \delta \cdot x(P)$$

is extreme, too.

(ii) Let x be a $b_s(x)$-optimal transportation vector, $i_0 \in U_s(x)$ and let P be a shortest x-augmenting path connecting i_0 with a node $j_0 \in U_t(x)$. Then for any $\delta \in \{1, \ldots, \kappa(P)\}$ the transportation vector

$$y := x \oplus \delta \cdot x(P)$$

is $b_s(y)$-optimal.

(iii)   Let x be a k-optimal transportation vector and let P be a shortest x-augmenting path. Then for any $\delta \in \{1, \ldots, \kappa(P)\}$ the transportation vector

$$y := x \oplus \delta \cdot x(P)$$

is $(k + \delta)$-optimal again.

Now assume an extreme transportation vector $\bar{x}$, then $\bar{x}$ is an optimal solution of the modified transportation problem

$$min \sum_{\{i,j\} \in E} c_{ij} x_{ij} \quad \text{subject to}$$

$$\sum_{j \in N(i)} x_{ij} = b_i(\bar{x}) \text{ for } i \in V_s$$

$$\sum_{i \in N(j)} x_{ij} = b_j(\bar{x}) \text{ for } i \in V_t$$

$$x_{ij} \geq 0 \text{ for } \{i, j\} \in E.$$

Hence exists a dual solution $(\bar{u}, \bar{v})$ with the properties

$$u_i + v_j \leq c_{ij} \quad \text{for } \{i, j\} \in E$$

$$u_i + v_j = c_{ij} \quad \text{if } \bar{x}_{ij} > 0.$$

Again we will call the pair $\bar{x}$ and $(\bar{u}, \bar{v})$ a compatible pair and consider the reduced costs $\bar{c}_{ij} = c_{ij} - u_i - v_j$ for $\{i, j\} \in E$.

Then for an augmenting path P connecting $i_0 \in V_s$ and $j_0 \in V_t$, the following relation holds

$$\bar{c}(P) = c(P) - u_{i_0} - v_{j_0}.$$

Thus when checking for the shortest augmenting path connecting a node $i_0 \in U_s(\bar{x})$ with a node $j_0 \in U_t(\bar{x})$ the reduced costs may be used instead of the original costs. Since the reduced costs are nonnegative, we may use the efficient Dijkstra-method here, too.

As for AP we can distinguish two versions of the SAP-labeling method

- the *tree–version*, where a single alternating tree is grown from a pre-specified node $i_0 \in U_s(\bar{x})$

- the *forest–version*, where simultaneously alternating trees are grown from all nodes in $U_s(\bar{x})$.

Then the labeling method for HTP differs from the labeling method for AP in two points, only:

(i) If $j^* \in L$ with $d_j^* = min\{d_j \mid j \in L\}$ has been found in <u>Step 1</u> and $b_{j^*}(\bar{x}) = b_{j^*}$ holds, then in <u>Step 2</u> <u>all</u> nodes $i^* \in V_s$ with $\bar{x}_{i^* j^*} > 0$ are labeled and then scanned.

(ii) If $j^* \in L$ with $d_j^* = min\{d_j \mid j \in L\}$ has been found in <u>Step 1</u> and $b_{j^*}(\bar{x}) < b_{j^*}$ holds, then a shortest augmenting path P has been found and its capacity $\kappa(P)$ has to be calculated.

Then after the augmentation $\tilde{x} := \bar{x} \oplus \kappa(P) \cdot \bar{x}(P)$ the same dual update as in the AP–case will produce a new compatible dual solution.

As for AP we can consider special refinements of the basic SAP–approach where a sequence of $b(x)$–optimal resp. k–optimal transportation vectors x is constructed. Since these refinements work analogously to the AP–procedures we do not go into detail here.

Finally, let us discuss the complexity of this approach. It is easy to see that transforming HTP into an equivalent AP over a graph G' and then applying the SAP–approach to the assignment problem on G' gives a procedure of complexity $O(\sum_{i \in V_s} b_i \cdot |E|)$. Any transportation vector x in G induces a matching $M(x)$ in G' with $|M(x)| = b(x)$ and any x–augmenting path P in G induces $\kappa(P)$ "parallel" $M(x)$–augmenting paths in G' . Thus let $P^{(1)}, \ldots, P^{(q)}$ be the sequence of augmenting paths constructed in the course of an SAP–application to an HTP on G then the following identity holds

$$q = \sum_{i \in V_s} b_i - \sum_{l=1}^{q} (\kappa(P^{(l)}) - 1).$$

Hence $\sum_{l=1}^{q}(\kappa(P^{(l)})-1)$ can be viewed as a measure of the "advantage" of applying the SAP–approach on G directly compared to the transformation to an assignment problem. In a sense we can call these $\sum_{l=1}^{q}(\kappa(P^{(l)})-1)$ augmentations "easy" or "free". Yet it can be shown that in the worst case $q = O(\sum_{i \in V_s} b_i)$ and thus the SAP–method for HTP is not "better" than the transformation approach and especially it is not a polynomial algorithm.

To obtain a polynomial algorithm for solving HTP the simple "greedy–philosophy" always to augment along a shortest augmenting path P has to be combined with a strategy which in some sense tries to "maximize" the number of "free" augmentations. Such an approach will be presented in the next section.

11.2. The Scaling Approach

In the last section we have interpreted the SAP–approach for HTP as a special implementation of the SAP–approach to the equivalent AP where several single augmentations are comprised into one general augmentation. Given an augmenting path P with capacity $\kappa(P)$ we have called $\kappa(P)-1$ of the $\kappa(P)$ single augmentations "free" augmentations or "easy" augmentations. Yet we have also seen that this greedy–approach of always performing the maximal possible number of $\kappa(P) - 1$ free augmentations does not improve the computational complexity of the SAP–approach.

Edmonds and Karp [1972] have proposed to perform a different type of "easy" augmentations which makes the SAP–approach for HTP a good algorithm. The essential idea of their so–called *scaling approach* is extremely simple:

Let x be an extreme transportation vector and $x_{i_0 j_0} > 0$ with $i_0 \in U_s(x)$

178

and $j_0 \in U_t(x)$ then $P = (V(P), E(P))$ with $V(P) = \{i_0, j_0\}$ and $E(P) = \{\{i_0, j_0\}\}$ is a shortest x–augmenting path connecting i_0 and j_0 since $\bar{c}_{i_0 j_0} = 0$ holds. Hence the transportation vector $\bar{x}$ with

$$\bar{x}_{ij} := \begin{cases} x_{i_0 j_0} + 1, & \text{for } \{i, j\} = \{i_0, j_0\} \\ x_{ij}, & \text{for } \{i, j\} \neq \{i_0, j_0\} \end{cases}$$

is an extreme transportation vector. Moreover if x and (u, v) are compatible then $\bar{x}$ and (u, v) are compatible, too. This obvious observation can easily be generalized to the following central theorem for the scaling approach:

Theorem 11.2.   *(Edmonds and Karp [1972])*
Let x be an extreme transportation vector and (u, v) a compatible dual solution. If $\bar{x}$ defined by $\bar{x}_{ij} := 2 \cdot x_{ij}$ for $\{i, j\} \in E$ is a transportation vector, then $\bar{x}$ is extreme and (u, v) is a compatible dual solution.

Now the $b(x)$ single augmentations when augmenting x to $\bar{x}$ $(:= 2 \cdot x)$ can be considered as "easy" augmentations since they can be performed at $O(|E|)$–cost all together.

Edmonds and Karp [1972] have given a simple strategy which in a sense maximizes the number of those easy augmentations and leaves only a polynomial number of non-easy augmentations to be performed, and leads to a good algorithm for solving HTP. It is an interesting fact that thereby the first type of easy augmentations coming from multiple augmentations along the "same" path are avoided completely.

Let $p \in \mathbb{N}$ be such that $2^{p-1} \leq max_{k \in V_s \cup V_t}\{b_k\} \leq 2^p$ then we define node capacities

$$b_k^{(l)} = \lfloor b_k/2^l \rfloor \qquad l = 0, 1, \ldots, p-1, \ k \in V_s \cup V_t$$

where $\lfloor x \rfloor$ denotes the smallest integer less than or equal x .

Let us denote by $G^{(l)}$ the graph $G = (V_s, V_t, E)$ with these altered node capacities. Then a vector $x : E \rightarrow \mathbb{N} \cup \{0\}$ is called a l-feasible transportation vector if

$$\sum_{j \in N(i)} x_{ij} \leq b_i^{(l)} \quad \text{for } i \in V_s$$

$$\sum_{i \in N(j)} x_{ij} \leq b_j^{(l)} \quad \text{for } j \in V_t.$$

An extreme l-feasible transportation vector x is called l-optimal if $b(x) = min\{\sum_{i \in V_s} b_i^{(l)}, \sum_{j \in V_t} b_j^{(l)}\}$ holds. Note that an extreme 0-feasible transportation vector $x^{(0)}$ is an optimal solution for the original problem.

Theorem 11.3.

Let $x^{(l)}$ be a l-optimal transportation vector and $(u^{(l)}, v^{(l)})$ a compatible dual vector. Then

(i) $y^{(l-1)} := 2 \cdot x^{(l)}$ is an extreme $(l-1)$- feasible transportation vector and $(u^{(l)}, v^{(l)})$ is a compatible dual solution.

(ii) A $(l-1)$-optimal transportation vector $x^{(l-1)}$ can be found starting from the compatible pair $y^{(l-1)}$ and $(u^{(l-1)}, v^{(l-1)})$ by at most $max\{|V_s|, |V_t|\}$ applications of the SAP-labeling method.

Proof. (i) follows immediately from theorem 11.2. since $y^{(l-1)}$ is $(l-1)$-feasible.

(ii): Let w.l.o.g. $b(x^{(l)}) = \sum_{i \in V_s} b_i^{(l)}$, then for any $i \in V_s$, $b_i^{(l-1)} - b_i(y^{(l-1)}) \in \{0, 1\}$. Hence at most $|V_s|$ augmentations are necessary in $G^{(l-1)}$.

$\square$

Thus starting with $y^{(p-1)} \equiv 0$ (and $(u^{(p-1)}, v^{(p-1)}) \equiv 0$) the total number K of augmentations to be performed by the shortest augmenting path method can be calculated as follows

$$K = b(x^{(p-1)}) + \sum_{l=1}^{p-1}(b(x^{(l-1)}) - b(y^{(l)}))$$

$$\leq b(x^{(p-1)}) + \sum_{l=1}^{p-1} max\{|V_s|, |V_t|\}$$

$$\leq p \cdot max\{V_s, V_t\}.$$

Now each shortest augmenting path computation and augmentation requires $O(|E|)$ computation steps. Thus this strategy leads to an algorithm the number of computation steps is bounded by a polynomial in the size of the problem, as measured by the length of the input.

We have introduced the scaling method as a special strategy within the general SAP–concept. A different interpretation was given by Edmonds and Karp. They viewed their approach as a method where sequentially HTP's of increasing "precision" are solved.

Since any MCFP can be reduced to HTP by a polynomial reduction (cf. section 5.1.) applying Edmonds and Karp's scaling technique to the equivalent HTP is a good algorithm for MCFP. Edmonds and Karp have shown how the scaling technique can be applied to MCFP directly. We do not discuss this extension here and refer to the original paper of Edmonds and Karp [1972]. In section 15.2. we will show how this basic idea of scaling the node–capacities leads to a polynomial algorithm for the b–matching problem, too.

Finally we want to mention that the existence of a HTP-algorithm which is polynomial in the size of the graph is still an open question.

Since the development of the "ellipsoid–method" and the proof that general linear programs can be solved in polynomial time the theoretical content of Edmonds and Karp's scaling method is subsumed by this more general result. Yet from a practical point of view the "ellipsoid–method" is not a serious candidate for an "efficient" HTP algorithm (or MCFP–algorithm). In this context

181

it is then an interesting observation, that a special refinement of the shortest-augmenting path concept leads to a good algorithm.

PART V

THE 1–MATCHING PROBLEM

In this part we discuss the key–problem within the class of general matching problems: the 1–matching problem (MP) in nonbipartite graphs, i.e.

$$min \ \{c(M) \mid M \ \text{perfect matching in} \ G\} \ .$$

Several authors treat the maximization problem instead (cf. Cunningham and Marsh [1978]) resp. allow even nonperfect matchings

$$max \ \{c(M) \mid M \ \text{matching in} \ G\}$$

(cf. Edmonds [1965], Lawler [1976]).

All these problems are equivalent in the sense that given a problem of one type it can be transformed into a problem of another kind such that the optimal solutions are the same.

Although Edmonds [1965] originally solved the max–weight matching problem we prefer to treat the min–cost perfect matching problem as the standard matching problem. We have already demonstated that all problems within the class of general matching problems can be reduced to an equivalent 1MP and from that fact we have argued that MP "contains" already the whole complexity of the entire class.

As for the bipartite case we will start with a discussion of the cardinality matching problem (CMP) in non–bipartite graphs. This is a purely combinatorial problem and the techniques and theorems developed for CMP will be used extensively in algorithmic approaches to the weighted problem.

Thereafter we will establish the so–called matching–polytope, i.e. the system of linear inequalities which describe the linear hull of matching vectors. The knowledge of this system is essential to be able to apply linear programming theory and LP–techniques to 1MP.

The following sections are then devoted to the different algorithmic approaches to 1MP

- the primal–dual blossom algorithm (Edmonds [1965])

- the primal method (Cunningham and Marsh [1978])

- the matching–simplex method (Edmonds and Koch [1981])

- the shortest augmenting path method (Derigs [1981])

In a separate section we will then again analyse the "near–equivalence" of matching algorithms.

Chapter 12. The Cardinality Matching Problem

In this chapter we discuss the problem of finding a matching of maximal cardinality in a general (nonbipartite) graph $G = (V, E)$. We will first introduce some combinatorial structures which typically arise in nonbipartite graphs and which have to be handled in cardinality matching algorithms. We also state the nonbipartite counterparts of the "existence theorems" and "characterizations" of perfect matchings in bipartite graphs. Then we present the alternating path labeling method for solving the cardinality matching problem in non–bipartite graphs.

12.1. Combinatorial Structures

Let $G = (V, E)$ be a (nonbipartite) graph. As in the bipartite case we denote by $\mathcal{M}$ the set of all matchings in G and by $\nu(G) = max \ \{|M| \mid M \in \mathcal{M}\}$ the matching number of G. Given $M \in \mathcal{M}$ the set of M–unsaturated nodes is denoted by $U(M)$ and M is called *perfect*, if $U(M) = \emptyset$. Perfect matchings are also called *1–factors* of G.

Now CMP can be formulated as integer program

$$max \ x(E) \qquad \text{subject to} \qquad \text{(CMP)}$$
$$x(\delta(v)) \leq 1 \quad \text{for } v \in V$$
$$x \in \{0,1\}^E \ .$$

Yet in contrast to the bipartite case, CMP can in general not be treated as linear program, i.e. the integrality condition is not superfluous, as the following example shows

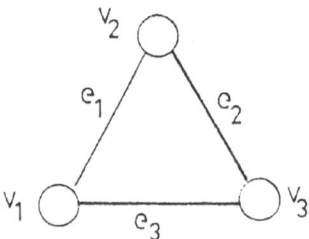

Figure 12.1. Counterexample K_3–graph

Obviously any maximal cardinality matching of K_3 contains exactly one edge while $x_{e_1} = x_{e_2} = x_{e_3} = \frac{1}{2}$ is an optimal LP–solution with $x(E) = \frac{3}{2}$.

While LP–instruments do not work for CMP in nonbipartite graphs the combinatorial tools and concepts introduced for bipartite matchings are applicable. So for instance Berge's theorem holds and the augmenting path method is applicable. Yet the labeling method presented in section 9.1. has to be modified, substantially.

Moreover the characterizations and existence–theorems for bipartite matchings and assignments are not valid here. This can also be seen from the example above where at least two nodes are needed to cover K_3 . Thus in nonbipartite graphs G the relation $\nu(G) < \tau(G)$ may hold. Those (nonbipartite) graphs G for which still $\nu(G) = \tau(G)$ holds are called *König–Egervary–graphs*. They have been studied and characterized in literature (cf. Deming [1979], Sterboul [1979]).

Consequently a different "stopping criterion" has to be developed for the augmenting path approach in general graphs. In the absence of such optimality conditions <u>all</u> alternating paths have to be enumerated which leads to an algorithm of exponentially growing amount. Although Berge [1972] reports, that such an enumerative labeling algorithm was able to solve "medium sized" problems within "reasonable" time, our own experience with the enumeration technique of Dörfler and Mühlbacher [1972], [1974] showed the limits of such approaches (cf. Derigs and Heske [1980]).

In fact Edmonds' famous blossom–algorithm for solving CMP (cf. Edmonds [1965]) can be viewed as a procedure where by means of "shrinking" certain subgraphs the original graph G is transformed into a König–Egervary–graph $\tilde{G}$ and the optimality of a matching in G then follows from the by König's theorem easy to check optimality of the induced matching in $\tilde{G}$.

In the following we will therefore first state theorems of Tutte and Edmonds which substitute the results of Hall and König in the nonbipartite case. Then we will introduce the concept of hypomatchable sets before we present Edmonds' CMP–method with its variants and improvements.

The first theorem, due to Tutte [1947], will play the role of Hall's theorem in the general case. We call a component of a graph *odd* if its number of nodes is odd.

Theorem 12.1.   *("Tutte's 1–factor theorem")*
A graph $G = (V, E)$ has a perfect matching iff $G - X$ has at most $|X|$ odd components for all $X \subseteq V$.

The condition is obviously necessary. A short and elegant proof of the sufficiency was given by Lovasz [1975] . Berge [1958] derived from Tutte's theorem the following formula for the matching number $\nu(G)$ of a general graph G :

Corollary 12.2.
For any graph $G = (V, E)$ the matching number is
$$\nu(G) = \min_{W \subset V} \frac{|V| + |W| - O(V \backslash W)}{2}$$
where $O(V \backslash W)$ is the number of odd components of the subgraph induced by $V \backslash W$.

Now let $\mathcal{N} = \{N_1, N_2, \ldots, N_q\}$ be a family of subsets of nodes with the property that $|N_i|$ is odd for $1 \leq i \leq q$. If $N_i = \{v\}$ with $v \in V$, we say that N_i covers $\delta(v)$ and we define $\mathrm{cap}(N_i) := 1$. If $|N_i| \geq 3$, we say that N_i covers $\gamma(N_i)$ and we define $\mathrm{cap}(N_i) := \frac{1}{2}(|N_i| - 1)$.

Now $\mathcal{N}$ is said to be an *odd–set–cover* of G if every edge $e \in E$ is covered by at least one set $N \in \mathcal{N}$ and the capacity of an odd–set–cover $\mathcal{N}$ is defined as

$$\mathrm{cap}(\mathcal{N}) := \sum_{i=1}^{q} \mathrm{cap}(N_i) .$$

It is easy to see that for any matching M in G and any odd–set–cover $\mathcal{N}$ in G

$$|M| \leq \mathrm{cap}(\mathcal{N})$$

holds.

Hence the above min–max–relation can be used as a sufficient optimality criterion in a CMP–algorithm. Moreover, Edmonds [1965] has shown that equality holds for all graphs and thus this min–max relation is a necessary optimality condition, too. In fact Edmonds has shown the validity of this relation in a constructive manner by giving an algorithm which for any graph produces at the same time a matching M and an odd–set–cover $\mathcal{N}$ fulfilling equality. Hence König's theorem for bipartite graphs has the following counterpart.

Theorem 12.3. *(Edmonds [1965])*
For any graph $G = (V, E)$ the cardinality of a maximum cardinality matching equals the capacity of a minimal odd–set–cover.

We will demonstrate the correctness of this result when discussing Edmond's CMP–algorithm.

The essential feature of Edmonds' CMP–approach is the idea of *shrinking* certain node–induced subgraphs $G[W] = (W, \gamma(W))$. For that purpose let

$W \subset V$, then we define the graph $G \times W$ which we obtain by shrinking $G[W]$ to a so–called *pseudonode* w in the following way:

$G \times W = (V', E')$ with
$$V' := V \backslash W \cup \{w\} \quad (w \notin V)$$
$$E' := \{\{i, w\} \mid i \in V \backslash W, \text{ ex. } j \in W \text{ with } \{i, j\} \in \delta(W)\}$$
$$\cup \{\{i, j\} \in E \mid i, j \notin W\}$$

The subgraphs which have to be shrunk in CMP–algorithms are all of a special structure . We will study this structure in the following and motivate the shrinking step thereby.

A subset $W \subseteq V$ with cardinality $|W| = 2k + 1$, $k \in \mathbb{N}$, is called *hypomatchable*, if for every $v \in W$ there exists a matching $M_v \in \mathcal{M}$ with $|M_v| = |M_v \cap \gamma(W)| = k$ leaving node v isolated. Then M_v is called *near-perfect* with respect to W and (due to Edmonds) the subgraph $(W, \gamma(W))$ is called a *blossom* with respect to M_v .

For example odd–cycles define obviously hypomatchable sets and Edmonds and Pulleyblank [1974] have shown that odd–cycles form the "generating structure" for hypomatchable sets. As a consequence we can say already at this point that "hypomatchability" is a consequence of nonbipartiteness and the key–concept for overcoming the difficulties stemming from nonbipartiteness. One should also mention here that the concept of hypomatchable sets although inherent in Edmonds' first formulation of the blossom–algorithm was not explicitly formulated prior to the Edmonds–Pulleyblank–paper. Due to this fact the few textbooks treating CMP use rather clumsy and badly motivated definitions of the structures (blossoms) to be handled (shrunk) in nonbipartite matching algorithms. To our opinion the introduction of the concept of hypomatchable sets facilitates the formulation and eases the understanding of Edmonds' approach significantly.

A family $\mathcal{A} \subseteq 2^V$ is called a *nested family* iff

(i) $|A| \geq 3$ for all $A \in \mathcal{A}$ and

(ii) $A, B \in \mathcal{A}$ with $A \cap B \neq \emptyset$ and $A \neq B$ then $A \subset B$ or $B \subset A$ holds.

Let $\mathcal{A}$ be a nested family over a finite set V then $|\mathcal{A}| \leq |V| - 1$ holds. This can be shown using elementary induction (cf. Pulleyblank [1973]).
For $A \in \mathcal{A}$ let us define $\mathcal{A}[A] := \{B \in \mathcal{A} \mid B \subset A,\ B \neq A\}$ and $G[A]$ the subgraph of G induced by A. Then the following theorem holds:

Theorem 12.4.   *(Edmonds and Pulleyblank [1975])*
A set $W \subset V$ is hypomatchable iff exists a nested family $\mathcal{A} \subseteq 2^W$ with the property

(i) $W \in \mathcal{A}$

(ii) $G[A] \times \mathcal{A}[A]$ *is spanned by an odd cycle for all $A \in \mathcal{A}$.*

A nested family $\mathcal{A}$ with property (ii) is called a *shrinking family.*
Let us denote by $\mathcal{S} = \mathcal{S}(G)$ the set of all hypomatchable subsets of V.

A set $A \in \mathcal{A}$ is called a *maximal member* of $\mathcal{A}$ iff exists no $B \in \mathcal{A}$, $B \neq A$ with $A \subset B$. Obviously $\mathcal{A} \backslash \{A\}$ is again a shrinking family for every maximal member $A \in \mathcal{A}$.

Now let $\mathcal{A}$ be a shrinking family and let $\{A_1, \ldots, A_q\}$ be the set of maximal members of $\mathcal{A}$ then we define

$$G \times \mathcal{A} := (\ldots((G \times A_1) \times A_2) \ldots A_q) .$$

Note that for maximal sets $A_i \in \mathcal{A}$ the order in which these are shrunk is of no consequence and the maximal sets can be identified with the pseudonodes in $G \times \mathcal{A}$. Now let $M \in \mathcal{M}$ and A a (maximal) member of $\mathcal{A}$. If $|M \cap \gamma(A)| = \frac{1}{2}(|A| - 1)$ we call $(A, \gamma(A))$ an (outermost) *blossom* with respect to M . From

190

the definition of hypomatchable sets resp. the definition of a shrinking family it is evident that every matching $M_{\mathcal{A}}$ in $G \times \mathcal{A}$ can be extended to a matching M in G such that every $A \in \mathcal{A}$ is a blossom with respect to M by the following recursive procedure:

<u>Procedure 12.1.</u>   **Blossom–Expansion–Procedure**

Until $\mathcal{A} = \emptyset$

    choose a maximal member $A \in \mathcal{A}$ with associated pseudonode a in $G \times \mathcal{A}$

      if a is matched with respect to $M_{\mathcal{A}}$ let $\{i,j\} \in E$ with $i \in A$ and $j \notin A$ be the edge which induces the matching edge e^* incident with pseudonode a in $G \times \mathcal{A}$,

      if a is not matched with respect to $M_{\mathcal{A}}$ choose $i \in A$ arbitrarily and set $e^* := \emptyset$.

    Let M_i be the near perfect matching in A leaving node i isolated.

    Set $M_{\mathcal{A}} := (M_{\mathcal{A}} \backslash \{e^*\}) \cup M_i$ and

      $\mathcal{A} := \mathcal{A} \backslash \{A\}$.

It is evident that given a shrinking family $\mathcal{A}$ in G , any perfect matching $M_{\mathcal{A}}$ in $G \times \mathcal{A}$ induces a perfect (and thus maximum cardinality) matching in G .

The following theorem is fundamental for Edmonds' blossom–algorithm

Theorem 12.5.   *(Edmonds* [1965]*)*

Let M be a matching in G and let $\mathcal{A}$ be a shrinking family in G such that every $A \in \mathcal{A}$ is a blossom with respect to M . Let $M_{\mathcal{A}}$ be the matching in $G \times \mathcal{A}$

which is induced by M . *Then every augmenting path* $P_{\mathcal{A}}$ *in* $G \times \mathcal{A}$ *with respect to* $M_{\mathcal{A}}$ *induces an augmenting path* P *with respect to* M *in* G .

Proof. Can be seen easily by successively applying the blossom–expansion–procedure to the maximal members $A \in \mathcal{A}$ the pseudonodes of which are contained in $P_{\mathcal{A}}$.

$\square$

We are now able to introduce Edmonds' CMP–approach i.e. the method of efficiently constructing augmenting paths in non–bipartite graphs.

12.2. The Blossom Algorithm

As for CMP in bipartite graphs alternating trees are grown from unsaturated nodes and again we can distinguish a *tree–version*, where a single tree is grown and a *forest–version*, where alternating trees from all unsaturated nodes are grown simultaneously. We will first introduce one phase of the tree–version here.

As in the bipartite case one starts growing a M–alternating tree rooted at a M–unsaturated node $r \in V$ using the rules intoduced in section 9.2. which partition the nodes of the tree $T = (V(T), E(T))$ into two classes $O(T)$, the set of outer nodes and $I(T)$, the set of inner nodes. Now these tree growing activities may be thwarted by the occurence of <u>case 4</u>:
"Branching out from a node $v \in O(T)$ to an edge $\{v, w\} \notin E(T)$ a node $w \in O(T)$ is reached."
This case indicates the existence of an odd cycle $C = (V(C), E(C))$ which can be identified in the following way. Let $b \in V$ be the first common node contained in the alternating paths from v to r resp. from w to r . Then from the alternating tree property it follows that $b \in O(T)$ and both paths are of even length. Now the odd cycle C is composed from P_{vb}, P_{wb} and edge $\{v, w\}$. Obviously M is

a near perfect matching of $V(C)$ and hence $(V(C), \gamma(V(C)))$ is a blossom with respect to M .

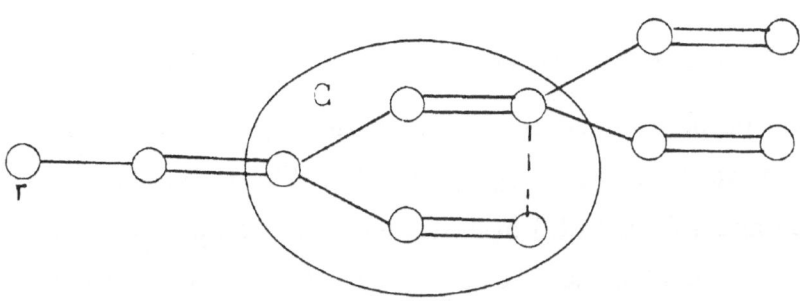

Figure 12.2. Alternating tree with blossom

Now the subgraph $G[V(C)]$ is shrunk to a pseudonode v_C . Let M' be the matching in $G \times V(C)$ which is induced by M .

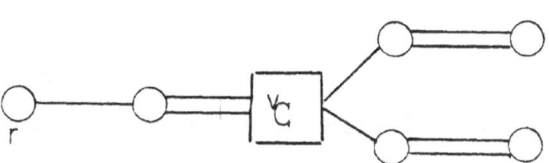

Figure 12.3. Alternating tree after shrinking the blossom

The M-alternating tree T induces a M'-alternating tree T' in $G \times V(C)$ with v_C an outer node of T' and we continue to grow the alternating tree T' in $G \times V(C)$.

During the course of the procedure the "shrink operation" may be applied a number of times. In particular, the shrink operation may be invoked recursively, in the sense that the set $V(C)$ may contain pseudonodes. Yet due to the construction, throughout the procedure we work on a graph $G \times \mathcal{A}$ with $\mathcal{A}$ a shrinking family and $M_{\mathcal{A}}$ the matching in $G \times \mathcal{A}$ which is induced by M. Thus from the above results it follows that any $M_{\mathcal{A}}$–augmenting path found in the course of this procedure induces a M–augmenting path in G. It remains to show that in the case that the alternating tree T' becomes a Hungarian tree in $G \times \mathcal{A}$, no M–augmenting path containing root r in G exists.

For that purpose we assume that after the occurence of a Hungarian tree T' in $G \times \mathcal{A}$ we delete the subgraph $(V(T'), E(T'))$ and the edges incident with (inner) nodes from $V(T')$ from further inspection and we choose another unsaturated node in G to start the next phase of the procedure. By this iterative process we end up with a graph $G \times \mathcal{A}'$, a matching $M_{\mathcal{A}'}$ in $G \times \mathcal{A}'$ and a collection of Hungarian trees in $G \times \mathcal{A}'$. Thus $M_{\mathcal{A}'}$ is a maximum cardinality matching in $G \times \mathcal{A}'$ where $\mathcal{A}'$ is a shrinking family in G. Now the following theorem which is due to Edmonds [1965], shows the optimality of the matching which is induced by $M_{\mathcal{A}'}$.

Theorem 12.6.

Let $\mathcal{A}$ be a shrinking family in $G = (V, E)$ and $M_{\mathcal{A}}$ a maximum cardinality matching in $G \times \mathcal{A}$. If either M is perfect or every pseudonode in $G \times \mathcal{A}$ which is contained in a Hungarian tree with respect to $M_{\mathcal{A}}$ is an outer node then $M_{\mathcal{A}}$ induces a maximum cardinality matching M in G.

Proof. If M is perfect the theorem is obviously true. Otherwise we construct an odd set cover $\mathcal{N}$ in G from the Hungarian trees in $G \times \mathcal{A}$ with respect to $M_{\mathcal{A}}$ in the following way:

Every inner node of a Hungarian tree becomes a singleton set in $\mathcal{N}$ and all hypomatchable sets corresponding to outer pseudonodes of the Hungarian trees become odd sets in the odd–set–cover $\mathcal{N}$, too. Finally we consider the (pseudo-) nodes in $G \times \mathcal{A}$ which are not contained in any Hungarian tree. If this set is void the odd–set–cover is complete. Otherwise associated with these nodes is a set $U \subset V$ of even cardinality. If $|U| = 2$, we make one of the two nodes a singleton set to complete $\mathcal{N}$. Otherwise we choose $u \in U$ arbitrarily and introduce $U' = U \backslash \{u\}$ and $\{u\}$ as sets into $\mathcal{N}$.

In any case $\mathcal{N}$ is an odd–set–cover and $\text{cap}(\mathcal{N}) = |M|$. Hence M is a maximum cardinality matching.

□

Since Edmonds' procedure always ends with the above constellation and hence an odd–set–cover of capacity equal to the cardinality of the current matching is always constructed the validity of theorem 12.3. is shown, too.

Note that the condition that all pseudonodes are <u>outer</u> nodes is essential for the validity of the theorem as the following example shows.

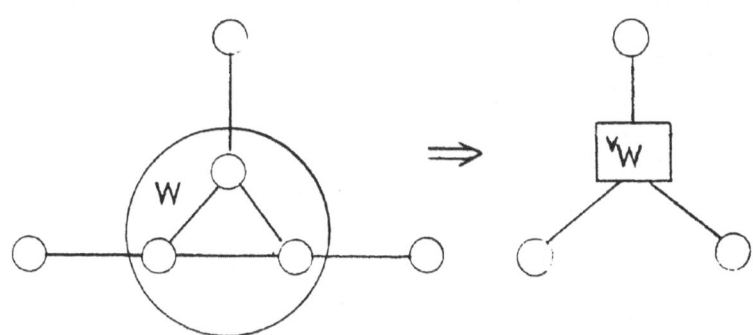

Figure 12.4. Counterexample

Here the set W is hypomatchable, yet shrinking $G[W]$ to a pseudonode leads to a graph $G \times W$ with the maximum cardinality matching M_W having cardinality 1 and any induced matching M in G having cardinality 2 while a maximum cardinality matching in G has cardinality 3.

As already mentioned, Edmonds' procedure can be interpreted as transforming the given graph G by shrinking certain hypomatchable sets into a graph $G \times \mathcal{A}$ which has a node cover not containing any pseudonode. In that case an odd–set–cover of G can be composed from the node cover in $G \times \mathcal{A}$ and the maximal members in $\mathcal{A}$.

In the *alternating forest* variant we would start growing alternating trees from all unsaturated nodes $r \in V$ simultaneously. Then case 4 may occur with two different implications:

Case (4a): The two outer nodes are contained in the same alternating tree.

Case (4b): The two outer nodes are contained in different alternating trees.

Case (4a) indicates the existence of an odd cycle / hypomatchable set to be shrunk while case (4b) indicates the existence of an augmenting path connecting the root–nodes of the two alternating trees.

After describing Edmonds' algorithm more verbally so far, in the following we formulate a labeling method for solving the phase subproblem of the alternating forest version of Edmonds' algorithm for solving CMP in general graphs. This labeling procedure calls for three "subroutines" which we describe now. Let s be an outer node of an alternating tree.

Given t, an outer node of the same alternating tree, the subroutine SHRINK (s, t) shrinks the subgraph $C = (V(C), \gamma(V(C)))$ spanned by the odd cycle induced by the edge $\{s, t\}$ to a pseudonode v_C . This pseudonode is made an outer node of the alternating tree in $G \times V(C)$.

Given t, an outer node of a different alternating tree, the subroutine AUGMENT (s, t) augments the matching M along the augmenting path P connecting the root–nodes of the two trees, i.e. $M := M \oplus P$. Then it expands all pseudonodes in both trees and extends the matching onto the hypomatchable sets. With this subroutine the current phase ends.

Given t an unlabeled node, i.e. a node not contained in any alternating tree, matched with the node w say, the subroutine GROW (s, t) labels node t "inner" and makes node w an "outer" node of the tree containing node s .

Procedure 12.2. Alternating forest labeling method for solving CMP

Step 0: Label all nodes $v \in U(M)$ "outer".

            If $U(M) = \emptyset$, Stop: M is perfect.

Step 1: Choose unscanned outer node $s \in V$.

            If no such node exists, Stop: M is of maximal cardinality.

            Otherwise given such a node s , goto Step 2.

Step 2: Scan node s in the following way

            for all nodes $t \in N(s)$

                if t outer node of the same tree as s $\rightarrow$ SHRINK (s,t)

                if t outer node of a different tree as s $\rightarrow$ AUGMENT (s,t)

                if t not labeled $\rightarrow$ GROW (s,t)

            goto Step 1.

In this labeling procedure the order in which the unscanned outer nodes are scanned is not prespecified. In fact special choice rules can obviously make this procedure grow only one alternating tree after another which would make it equivalent to the alternating tree version. A common rule is the first–labeled–first–scanned principle which leads to a breadth–first–search of the graph.

There are variants where after the successful end of a phase i.e. an augmentation, all labels on nodes not contained in the two trees leading to the augmenting path are kept for the next phase. Yet our own experience showed no advantage for this procedure.

Edmonds did not explicitly use the concept of hypomatchable sets when first introducing his blossom–algorithm, which he called a " conceptual description of an algorithm" rather than an algorithm and he claimed that his method could be implemented in order $O(|V|^4)$. The labeling procedure given above leaves the question open how to implement the three subroutines and how to backtrack alternating paths when checking for a blossom or an augmenting path. Here several implementations using different data–structures have been developed.

In fact it took about a decade until proper labeling techniques and data–structures were developed which enabled the implementation of Edmonds' method (cf. Gabow [1975], Lawler [1976]). Immediately after Edmonds' pioneering article, Witzgall and Zahn [1965] published a "modification of Edmonds' algorithm" which avoided shrinking of subgraphs by special labeling techniques. Yet they did not discuss the computational complexity of their approach. Gabow's implementation — an ALGOL–code of which has been published — is of order $O(|V|^3)$ and makes use of a special adjacency list structure for storing the graph which has been developed by Tarjan. As Gabow noted, the running time can be reduced to $O(|V||E| \cdot \alpha(|V|, |E|))$ where α is the functional inverse of Ackermann's function (cf. Aho et. al. [1974]). Recently Gabow and Tarjan [1983] developed a linear time $SET-UNION - algorithm$ which, incorporated into Gabow's implementation, reduces the complexity to $O(|E| \cdot |V|)$. Earlier Heske [1978] has already developed an $O(|V| \cdot |E|)$–modification of Gabow's labeling method a FORTRAN–code of which is published in Derigs and Heske [1980]. The use of Tarjan's data–structure is not essential for Gabow's implementation and seems to be quite ineconomical.

Recently a FORTRAN–code for solving CMP has been published by Conradt and Pape [1980] resp. Conradt [1978]. They use an extremely simple labeling technique for growing single alternating trees via a breadth–first–search. Instead of shrinking a blossom they develop the odd cycle into two alternating paths. They did not discuss the complexity of this approach but computational results are presented in Conradt [1978] which show that the method is superior to Gabow's method.

We developed two efficient implementations of Edmonds' method based on the alternating tree version resp. the alternating forest version. These implementations use a special "recursive list" for storing the nodes contained in hypomatchable sets which are shrunk during the process. The alternating forest version was shown to be superior to all the other methods mentioned above. A FORTRAN–code of this procedure is listed in Burkard and Derigs [1980].

The nontrivial adaption of Hopcroft and Karp's "multi–augmentation" approach to the nonbipartite case was first accomplished by Even and Kariv [1975]. In Kariv [1976] different implementations are presented which lead to routines of complexity $O(|V|^{\frac{5}{2}})$, $O(|V|^{\frac{1}{2}} \cdot |E| \cdot lg\, lg\, |V|)$ and $O(|V|^{\frac{1}{2}} \cdot |E| + |V|^{\frac{3}{2}+\epsilon})$ with $\epsilon > 0$ arbitrarily, depending on the data–structure which is used. For this reduction of computational complexity one has to pay a considerable amount of extra storage capacity. In Derigs and Heske [1980] it is shown that even for large graphs a FORTRAN–version of Kariv's original PL1–code is inferior with respect to running time compared to codes based on Gabow's implementation.

While Even and Kariv's implementation is rather involved, recently, Micali and Vazirani [1980] , using the same basic idea, obtained a simplified algorithm of complexity $O(|V|^{\frac{1}{2}} |E|)$. To our knowledge this approach was never coded.

Chapter 13. The Min–Cost Perfect Matching Problem

In this chapter we discuss the problem of constructing a min–cost perfect matching in general graphs. We have introduced 1MP as an integer program using the node–edge incidence matrix A . In the bipartite case we could show that A is totally unimodular and hence AP can be solved by solving the LP–relaxation of the integer program. This approach is not possible for nonbipartite graphs since the polytope associated with the LP–relaxation may have fractional vertices. Thus before discussing different algorithmic principles, we first establish the so–called *matching polytope*, i.e. the polyhedron the vertices of which correspond to matchings in G .

13.1. The Matching Polytope

Let $G = (V, E)$ be a graph and w.l.o.g. assume $|V|$ even. Then we associate with every matching M in $\mathcal{M}$ an incidence vector $x_M \in \mathbb{R}^E$ by setting

$$x_M^e = \begin{cases} 1 & \text{if } e \in M \\ 0 & \text{if } e \notin M . \end{cases}$$

Then we know that the set $X_{\mathcal{M}} \subset \{0,1\}^E$ of "matching vectors" is given by the system

$$Ax \le 1$$

$$x \ge 0 , \quad \text{integer valued,}$$

with A the node–edge incidence matrix of G .

Now $P_{\mathcal{M}} := \text{conv}(X_{\mathcal{M}}) = \text{conv}\{x_M \in \mathbb{R}^E \mid M \in \mathcal{M}\}$ is called the *monotone matching polytope* of G while $P_{\mathcal{M}_P} := \text{conv}\{x_M \in \mathbb{R}^E \mid M \in \mathcal{M}_P\}$ is called the *perfect matching polytope* of G . $P_{\mathcal{M}}$ and $P_{\mathcal{M}_P}$ are subsets of the unit hypercube in $\mathbb{R}^E$ and every vertex of $P_{\mathcal{M}}(P_{\mathcal{M}_P})$ corresponds to a (perfect) matching in G . Thus given a mapping (cost–function) $c : E \to \mathbb{R}$ the 1–matching problem

$$min \; \{c(M) \mid M \in \mathcal{M}_P\}$$

can be solved by solving the problem

$$min \; \{c'x \mid x \in P_{\mathcal{M}_P}\} \; .$$

The (perfect) matching polytope has been studied by several authors. The first representation of $P_{\mathcal{M}}(P_{\mathcal{M}_P})$ in terms of a (finite) system of linear inequalities and equations was given by Edmonds [1965] who proved the validity of the LP–system in an algorithmic way by giving a combinatorially natured algorithm for solving the matching problem which makes use of and thereby proofs the duality relationship associated with the LP–system.

A rather involved inductive proof was later given by Balinski [1972], while recently an elegant short proof for the characterization of $P_{\mathcal{M}_P}$ was given by Schrijver [1981].

It is easy to see that $P_{\mathcal{M}}$ contains $|E|+1$ affinely independent matching vectors, hence $P_{\mathcal{M}}$ is full dimensional. Thus a unique complete and nonredundant system of inequalities describing $P_{\mathcal{M}}$ exists. These irredundant (in)equalities are called the *facets* of $P_{\mathcal{M}}$. The system of facets for $P_{\mathcal{M}}$ was developed by Edmonds and Pulleyblank [1974].

In the following we will present the main results on $P_{\mathcal{M}}$ and $P_{\mathcal{M}_P}$ which are basic for all efficient MP–algorithms.

With A the node–edge incidence matrix of G we call the system

$$F_{\mathcal{M}} := \{x \in \mathbb{R}^E \mid Ax \leq 1, \; x \geq 0\} \supset P_{\mathcal{M}}$$

the *fractional matching polytope* and the extreme points of $F_{\mathcal{M}}$ are called *fractional matchings* of G . Note that $F_{\mathcal{M}_P}$ the *fractional perfect matching polytope* is defined analogously.

We have already demonstrated that $F_{\mathcal{M}}$ may have fractional vertices hence the *linear programming relaxation* of a matching problem

$$min \; \{c'x \mid x \in F_{\mathcal{M}}\}$$

may have a nonfeasible (i.e. non–integer) optimal solution.

The following theorem is due to Balinski [1968].

Theorem 13.1.

Any vertex x of $F_{\mathcal{M}}$ ($F_{\mathcal{M}_P}$) is $(0, 1, \frac{1}{2})$–valued. Let $E_{\frac{1}{2}} := \{e \in E \mid x_e = \frac{1}{2}\}$ and $G_{\frac{1}{2}} = (V, E_{\frac{1}{2}})$ then each component of $G_{\frac{1}{2}}$ is either an isolated node or a (cord–free) odd–cycle ("odd polygon").

Example

The following example shows the existence of non–integer extreme points for $\mathcal{F}_{\mathcal{M}_P}$. Let $G = (V, E)$ be the following graph

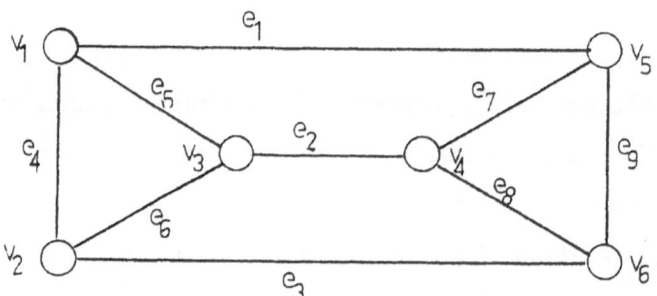

Figure 13.1.

Then $x_1 = x_2 = x_3 = 0$ implies $x_4 + x_5 = x_5 + x_6 = 1$ hence $x_4 = x_6$. Then from $x_4 + x_6 = 1$ we get $x_4 = x_6 = \frac{1}{2}$ etc. Thus $x_1^* = x_2^* = x_3^* = 0$, $x_4^* = x_5^* = x_6^* = x_7^* = x_8^* = x_9^* = \frac{1}{2}$ is an extreme point of $F_{\mathcal{M}_P}$.

From a theoretical and computational point of view it is an interesting fact that the linear programming relaxation of MP on $G = (V, E)$:

$$min \ \{c'x \mid x \in F_{\mathcal{M}_P}\}$$

is equivalent to an assignment problem in a related symmetric bipartite graph $G' = (V_1, V_2, E')$. Here V_1 and V_2 are two copies of the node set V and E' is constructed by introducing two edges $\{i_1, j_2\}$ and $\{j_1, i_2\}$ for every edge $\{i, j\} \in E$ and the mapping $c' : E' \rightarrow \mathbb{R}$ is given by $c'_{i_1 j_2} := c'_{j_1 i_2} := c_{ij}$ for $\{i, j\} \in E$.

Given an optimal assignment M' in G' with "assignment vector" x', an optimal fractional matching x is given by

$$x_{ij} = \frac{1}{2}(x'_{i_1 j_2} + x'_{j_1 i_2}) \ .$$

Thus x is a (non–fractional) matching iff x' is *symmetric* i.e.

$$x'_{i_1 j_2} = x'_{j_1 i_2} \quad \text{for all } \{i,j\} \in E \text{ resp.}$$
$$\{i_1, j_2\} \in M' \quad \Leftrightarrow \quad \{j_1, i_2\} \in M' \quad \text{holds} \ .$$

Hence 1MP on G is equivalent to the *symmetric assignment problem* on G' . This relation was used by Murty [1967] in a branch and bound–algorithm for solving 1MP. Computational results with this enumerative "matching approach" are given in Devine and Glover [1972] and Roberts [1969]. The convex hull of the symmetric assignments in a symmetric bipartite graph has been characterized by Cruse [1975].

Now let M be any matching in G and $W \subset V$ with $|W| \geq 3$, odd. Then the following relation is obviously fulfilled

$$|M \cap \gamma(W)| \leq (|W| - 1)/2.$$

Thus the inequality

$$x(\gamma(W)) \leq \frac{1}{2}(|W| - 1)$$

is fulfilled by any $x \in X_{\mathcal{M}}$.

Yet given a noninteger vertex $\bar{x}$ of $F_{\mathcal{M}}$ and $W \subseteq V$ a node–set which is spanning an odd–cycle in $G(\bar{x})$

$$\bar{x}(\gamma(w)) = \frac{1}{2}|W|$$

holds. Thus the above inequalities, which are called *blossom–inequalities* due to Edmonds [1965], are valid for $P_{\mathcal{M}}$ and "cut–off" all noninteger vertices of $F_{\mathcal{M}}$. Let us define $q_W := \frac{1}{2}(|W| - 1)$ for $W \subseteq V$.

For the above example the set $W = \{v_1, v_2, v_3\}$ induces the blossom–inequality

$$x_4 + x_5 + x_6 \leq 1$$

which is not fulfilled by the extreme solution $x^* \in F_{M_P}$.

Now Edmonds could show that the set of blossom constraints is sufficient to describe the matching polytope $P_{\mathcal{M}}$, i.e. given a graph $G = (V, E)$, the extreme points of the system

$$x(\delta(i)) \leq 1 \quad \text{for } i \in V$$

$$x(\gamma(W)) \leq q_W \quad \text{for } W \subset V, \ |W| \geq 3, \text{ odd}$$

$$x \ \geq \ 0$$

are integer valued.

Note that the number of blossom constraints grows exponentially with the size of the graph, i.e. the number of nodes. However this set of inequalities is generally far from minimal. The unique minimal subset of blossom-inequalities which serves to describe $P_{\mathcal{M}}$, i.e. the facets of $P_{\mathcal{M}}$, were characterized by Edmonds and Pulleyblank [1974].

A *cutnode* of a graph $G' = (V', E')$ is a node $w \in V'$ such that $G'[V' \setminus \{w\}]$ has more connected components than G . Let us call a graph G' *nonseparable* if it does not contain any cutnode. Then the following theorem holds

Theorem 13.2. *(Edmonds and Pulleyblank [1974])*
The blossom–inequality
$$x(\gamma(W)) \leq q_W$$

is a facet of $P_{\mathcal{M}}$ for $W \subset V$, $|W| \geq 3$, odd iff $G[W]$ is hypomatchable and nonseparable.

Thus in general the number of inequalities necessary to describe $P_\mathcal{M}$ may still be exponentially growing with the size of the graph. Yet the facets have an attractive combinatorial structure which will show to be convenient from a computational point of view. Let us denote by $Q = Q(G)$ the set of facet-inducing subsets of V i.e.

$$Q = \{W \subset V \mid |W| \geq 3 \,, \text{ odd}, \ G[W] \quad \text{hypomatchable and nonseparable .}\}$$

Note that with this definition $Q \subseteq S$.

Let us now turn to the linear description of $P_{\mathcal{M}_P}$. From the above results follows immediately that the following system of linear inequalities — if it is non–empty — has only integer–valued extreme points

$$x(\delta(i)) = 1 \quad \text{for } i \in V$$

$$x(\gamma(W)) \leq q_W \quad \text{for } W \in Q$$

$$x \ \geq \ 0$$

The linear characterization of $P_{\mathcal{M}_P}$ is not unique and we call the above system the *blossom–characterization* of $P_{\mathcal{M}_P}$ or characterization I .

Thus with the perfect matching problem

$$min \ \{c'x \mid x \in P_{\mathcal{M}_P}\}$$

we can associate the following dual (matching) problem

$$max \ \sum_{i \in V} y_i - \sum_{W \in Q} q_W \cdot y_W \quad \text{subject to} \qquad \text{(DMPI)}$$

$$y_i + y_j - \sum_{\{i,j\} \in \gamma(W)} y_W \leq c_{ij} \quad \text{for } \{i,j\} \in E$$

$$y_W \ \geq \ 0 \quad \text{for } W \in Q$$

and an optimal perfect matching M^* can be characterized by the existence of a complementary DMPI–feasible solution y, i.e.

$$\{i,j\} \in M^* \ \Rightarrow \ y_i + y_j - \sum_{\{i,j\} \in \gamma(W)} y_W = c_{ij}$$

$$y_W > 0 \ \Rightarrow \ |M^* \cap \gamma(W)| = q_W$$

An alternative to the above blossom–characterization of $P_{\mathcal{M}_P}$ is the following so–called *cut–characterization* where each blossom–inequality for $W \in \mathcal{Q}$ is replaced by an equivalent so–called *cut inequality*

$$x(\delta(W)) \geq 1 .$$

Thus the complete system (characterization II) is :

$$x(\delta(i)) = 1 \quad \text{for} \quad i \in V$$
$$x(\delta(W)) \geq 1 \quad \text{for} \quad W \in \mathcal{Q}$$
$$x \quad \geq \quad 0$$

and we get the following dual (matching) problem

$$max \; \sum_{i \in V} y_i + \sum_{W \in \mathcal{Q}} y_W \qquad \qquad \text{(DMPII)}$$
$$y_i + y_j + \sum_{\{i,j\} \in \delta(W)} y_W \leq c_{ij} \quad \text{for} \quad \{i,j\} \in E$$
$$y_W \geq 0 \quad \text{for} \quad W \in \mathcal{Q} .$$

Then the associated complementary slackness conditions characterizing an optimal $M^* \in \mathcal{M}_P$ are

$$\{i,j\} \in M^* \;\; \Rightarrow \;\; y_i + y_j + \sum_{\{i,j\} \in \delta(W)} y_W = c_{ij}$$
$$y_W > 0 \;\;\; \Rightarrow \;\; |M^* \cap \delta(W)| = 1 .$$

The equivalence of both linear descriptions follows immediately, since characterization II is obtained from characterization I by an invertible linear transformation:

For $W \in \mathcal{Q}$ and $x \in \mathbb{R}^E$ we have

$$x(\delta(W)) = \sum_{i \in W} x(\delta(i)) - 2x(\gamma(W)).$$

Thus let α_i, α_W denote the node– and blossom–inequalities from characterization I and let β_i, β_W denote the node– and cut–inequalities of characterization II, then

$$\beta_i = \alpha_i \qquad \text{for } i \in V$$

$$\beta_W = \sum_{i \in W} \alpha_i - 2\alpha_W \quad \text{for } W \in \mathcal{Q}.$$

Also there is an one–to–one correspondence between feasible solutions of the two dual programs. Given a feasible solution y for DMPI the associated feasible solution $\bar{y}$ for DMPII is given by

$$\bar{y}_i = y_i - \frac{1}{2} \sum_{i \in W} y_W \quad \text{for } i \in V$$

$$\bar{y}_W = \frac{1}{2} y_W \qquad \text{for } W \in \mathcal{Q}.$$

Inversely, given $\bar{y}$ feasible for DMPII the associated feasible solution y for DMPI is

$$y_i = \bar{y}_i + \sum_{i \in W} \bar{y}_W \quad \text{for } i \in V$$

$$y_W = 2\bar{y}_W \qquad \text{for } W \in \mathcal{Q}.$$

Moreover y is a basic (optimal) solution for DMPI iff the associated $\bar{y}$ is a basic (optimal) solution for DMPII.

In the next section we will present different approaches for solving 1MP. All these procedures use the knowledge of the linear characterization of $P_{\mathcal{M}_P}$, more precisely the associated complementary slackness conditions, extensively. Since there are two different characterizations available, in principle, each approach leads to two procedures based on the different characterizations.

Although the blossom–characterization is more widely used in literature and in implementations, to our opinion, the cut–characterization has some computational advantage.

Moreover the cut–characterization is more closely related to recent problems solved by Seymour [1979] and Padberg and Rao [1982] which we do not discuss in this work.

13.2. The Primal–Dual Method

In this section we discuss the celebrated (primal–dual) blossom–algorithm for solving MP which has been developed by Edmonds [1965]. In fact Edmonds did not treat MP but the related maximum–weight–matching problem

$$max \ \{c(M) \mid M \in \mathcal{M}\} \ .$$

Yet his results are immediately adaptable for MP.

In his paper Edmonds actually solved two major problems in one step — establishing the matching polytope and formulating a good matching algorithm. The simultaneous approach turned out to be a key idea for tackling combinatorial optimization problems. All authors describing the blossom–algorithm today are in a much better position – they use the knowledge on the matching polytope and the knowledge of the essential inequalities (facets) of the matching polytope. When studying the early "matching–papers" — which are often rather clumsy — and comparing their style of exposition one should have this in mind.

In the following we will

- introduce the blossom algorithm as a special implementation of the primal–dual–scheme for linear programs and

- exploit the combinatorial background of the subproblems to be solved in the course of the procedure

Thereby we will use characterization I (blossom–inequalities) of the matching polytope. The labeling algorithm based on characterization II (cut–inequalities) of the matching polytope will be given at the end of this section. Using the blossom–characterization 1MP has the following dual program:

$$max \sum_{i \in V} y_i + \sum_{R \in S} q_R \cdot y_R \quad \text{subject to} \qquad \text{(DMPI)}$$

$$y_i + y_j + \sum_{\{i,j\} \in \gamma(R)} y_R \leq c_{ij} \quad \text{for } \{i,j\} \in E$$

$$y_R \leq 0 \quad \text{for } R \in S.$$

Note that instead of Q we use the set S of hypomatchable sets here. This has only technical reasons, since we will apply the theory developed for CMP and the family S here. Moreover we have transformed (DMPI) into an equivalent form where the variables y_R are nonpositive. To develop the primal–dual mechanism it is more convenient to use this equivalent program.

Let $y \in \mathbb{R}^{V \cup S}$ be a feasible dual solution and let again the reduced cost of an edge $\{i,j\} \in E$ be denoted by $\bar{c}_{ij}$. As in the Hungarian method for solving AP we consider

$$E(y) := \{\{i,j\} \in E \mid \bar{c}_{ij} = 0\}$$

the set of *admissible edges* with respect to y and in addition

$$S(y) := \{R \in S \mid y_R = 0\}$$

the set of *"admissible" hypomatchable sets* with respect to y. Now a feasible dual solution is called *strongly dual feasible* iff $\overline{S(y)} := S \backslash S(y)$ is a shrinking family in $G(y) = (V, E(y))$.

For example $y \equiv 0$ is a strongly feasible dual solution and commonly used to start the blossom–algorithm.

In general the blossom algorithm may start from any strongly feasible dual solution y and it maintains strongly dual feasibility, throughout. With respect to a strongly dual feasible solution y we define the restricted primal problem as follows:

$$\min \sum_{i \in V} d_i \quad \text{subject to} \tag{RP}$$

$$x(\delta(i)) + d_i = 1 \quad \text{for } i \in V$$

$$x(\gamma(R)) \leq q_R \quad \text{for } R \in S$$

$$x(\gamma(R)) = q_R \quad \text{for } R \in \overline{S(y)}$$

$$x_{ij} \geq 0 \quad \text{for } \{i,j\} \in E$$

$$x_{ij} = 0 \quad \text{for } \{i,j\} \notin E(y)$$

$$d_i \geq 0 \quad \text{for } i \in V .$$

Now the dual of the above program is

$$\max \sum_{i \in V} y_i + \sum_{R \in S} q_R \cdot y_R \tag{DRP}$$

$$y_i + y_j + \sum_{\{i,j\} \in \gamma(R)} y_R \leq 0 \quad \text{for } \{i,j\} \in E(y)$$

$$y_R \leq 0 \quad \text{for } R \in S(y)$$

$$y_i \leq 1 \quad \text{for } i \in V$$

$$y_R \quad \text{unrestricted in sign for} \quad R \in \overline{S(y)}$$

Now it is easy to see that RP is the purely combinatorial problem of finding a maximum cardinality matching $\bar{M}$ in $G \times \overline{S(y)}$.

Let $\bar{M}$ be a maximum cardinality matching in $\bar{G} := G(y) \times \overline{S(y)}$. Then we have to consider two cases

<u>Case 1:</u> $\bar{M}$ is perfect in $\bar{G}$.

Then $\bar{M}$ induces a perfect matching M in G with the property that M and the dual solution y fulfill the complementary slackness conditions. Hence M is an optimal perfect matching in G .

<u>Case 2:</u> $\bar{M}$ is a nonperfect matching M in G .

In that case we construct a solution $\bar{y}$ for DRP having the same objective function value $|M|$. The CMP–labeling procedure has constructed the maximum cardinality matching in $\bar{G}$ by growing alternating trees and possibly shrinking certain hypomatchable sets $R \in S(y)$. Let $\tilde{S} \subseteq S$ be the set of hypomatchable sets which are shrunk at the end of the CMP–labeling method and $\bar{M}$ the maximum cardinality matching in $G(y) \times \tilde{S}$ which induces $\bar{M}$. Since $\overline{S(y)}$ was a shrinking family in $G(y)$, $\tilde{S}$ is a shrinking family in $G(y)$, too. Note that all newly shrunk hypomatchable sets $R \in \tilde{S} \backslash \overline{S(y)}$ have become <u>outer</u> pseudonodes of an alternating tree.

Then the set V of nodes and the set S of hypomatchable sets in G is partioned into three classes, each:
Let $\tilde{S}_0 (\tilde{S}_I)$ be the maximal sets in $\tilde{S}$ the pseudonode of which is an outer (inner) node of a Hungarian tree in $G(y) \times \tilde{S}$ with respect to $\bar{M}$ and let V_0 (V_I) be the set of nodes which are outer (inner) nodes of a Hungarian tree in $G(y) \times \tilde{S}$ with respect to $\bar{M}$ or contained in a set $R \in \tilde{S}_0$ ($R \in \tilde{S}_I$) .

Based on this partition define

$$
\bar{y}_i = \begin{cases} 1 & \text{for } i \in V_0 \\ -1 & \text{for } i \in V_I \\ 0 & \text{for } i \in V \backslash (V_0 \cup V_I) \end{cases}
$$

$$
\bar{y}_R = \begin{cases} -2 & \text{if } R \in S_0 \\ +2 & \text{if } R \in S_I \\ 0 & \text{if } R \in S \backslash (S_0 \cup S_I) \ . \end{cases}
$$

Then $\bar{y}$ is a feasible solution for DRP since $S_I \subset \overline{S(y)}$ holds.

Moreover the value of $\bar{y}$ can be calculated in the following way:

$$
\sum_{i \in V} \bar{y}_i + \sum_{R \in S} \frac{1}{2}(|R| - 1) \cdot \bar{y}_R = (|V_0| - \sum_{R \in S_0} (|R| - 1) - (|V_I| - \sum_{R \in S_I} (|R| - 1)) \ .
$$

But this value is equal to the difference between the number of outer labeled (pseudo–) nodes in $G(y) \times \tilde{S}$ and the number of inner labeled (pseudo-) nodes in $G \times \tilde{S}$. And this number equals the number of unmatched nodes in $G \times \tilde{S}$ which is equal to the number of unmatched nodes in G with respect to the (not necessarily maximum cardinality) matching M which is induced by $\tilde{M}$.

Now the optimal solution $\bar{y}$ of DRP is used to update the dual solution y in the following way:

First we have to investigate those edges $\{i,j\} \in E$ for which

$$\bar{y}_i + \bar{y}_j + \sum_{\{i,j\} \in \gamma(R)} \bar{y}_R > 0$$

holds.

This can happen in two cases:

<u>Case 1:</u> $i,j \in V_0$, yet not contained in a common hypomatchable set $R \in \tilde{S}$.

In that case $\bar{c}_{ij} = c_{ij} - y_i - y_j$ and $\bar{y}_i + \bar{y}_j + \sum_{\{i,j\} \in \gamma(R)} \bar{y}_R = 2$ holds.

<u>Case 2:</u> $i \in V_0$ and $j \in V \backslash (V_0 \cup V_I)$

In that case $\bar{c}_{ij} = c_{ij} - y_i - y_j$ and $\bar{y}_i + \bar{y}_j + \sum_{\{i,j\} \in \gamma(R)} \bar{y}_R = 1$ holds.

Thus we calculate

$$\Theta_1 := min \ \{\frac{1}{2}\bar{c}_{ij} \mid i,j \in V_0 \text{, not contained in a common}$$
$$\text{hypomatchable set } R \in \tilde{S}\}$$

$$\Theta_2 := min \ \{\bar{c}_{ij} \mid i \in V_0 \text{, } j \in V \backslash (V_0 \cup V_I)\} \ .$$

Second we have to discuss those hypomatchable sets $R \in S$ for which $\bar{y}_R > 0$. This is possible for $R \in S_I$ only, where $\bar{y}_R = 2$ holds. Hence we calculate

$$\Theta_3 := min \ \{\frac{-y_R}{2} \mid R \in \tilde{S}_I\} \ .$$

With $\Theta := min \ \{\Theta_1, \Theta_2, \Theta_3\}$ the "improved" dual solution $\tilde{y}$ is then given by:

$$\tilde{y}_i := \begin{cases} y_i + \Theta & \text{for } i \in V_0 \\ y_i - \Theta & \text{for } i \in V_I \\ y_i & \text{for } i \in V \backslash (V_I \cup V_0) \end{cases}$$

$$\tilde{y}_R := \begin{cases} y_R - 2\Theta & \text{for } R \in S_0 \\ y_R + 2\Theta & \text{for } R \in S_I \\ y_R & \text{for } i \in V \backslash (V_I \cup V_0) \end{cases}$$

This *dual–change* has the following consequences:

If $\Theta < \Theta_3$, the next RP is to find a maximum cardinality matching in $G(\tilde{y}) \times \tilde{S}$. Here the "old" maximum cardinality matching $\bar{M}$ and the Hungarian trees are "admissible" and thus can be used to initialize the CMP–labeling method.

If $\Theta = \Theta_2$, at least one Hungarian tree can be grown in $G(\tilde{y}) \times \tilde{S}$ and if $\Theta = \Theta_1$, either an augmenting path can be found or another blossom $R \in S \backslash \tilde{S}$ with respect to M is "created" in $G(\tilde{y}) \times \tilde{S}$ and will be shrunk.

If $\Theta = \Theta_3$, one maximal member R of $\tilde{S}$ has zero dual weight y_R , and thus need not kept shrunk furtheron. Hence we will generate the graph $G(\tilde{y}) \times (\tilde{S} \backslash \{R\})$ and extend the matching $\bar{M}$ to the odd cycle spanning $G(\tilde{y})[R] \times \tilde{S}[R]$. Then we will continue with the CMP–labeling technique in $G(\tilde{y}) \times (\tilde{S} \backslash \{R\})$.

Using the concept of a node–cover in $G(y) \times \tilde{S}$ resp. the concept of odd-set–covers in $G(y) \times \overline{S(y)}$ the primal dual algorithm can also be interpreted and motivated in a more combinatorial way as follows:

Starting from the graph $G(y) \times \overline{S(y)}$ the CMP–labeling procedure creates a König–Egervary graph $G(y) \times \tilde{S}$ by eventually shrinking some hypomatchable sets $R \in S(y)$. Then the inner (pseudo-) nodes in $G(y) \times \tilde{S}$ plus the non–tree nodes induce a node–cover in $G(y) \times \tilde{S}$.

Now there are two possibilities: the optimal perfect matching may be induced by a perfect matching in $G \times \overline{S(y)}$ or at least one hypomatchable set $R \in \overline{S(y)}$ will not be a blossom with respect to the optimal matching. In case 1 and case 2 ($\Theta < \Theta_3$) the "cheapest" edge not covered by the optimal cover in $G \times \overline{S(y)}$ is made admissible to eventually enable the construction of a perfect matching in $G(\tilde{y}) \times \overline{S(\tilde{y})}$ which is a subgraph of $G \times \overline{S(y)}$ .This matching then induces an optimal matching in G . Case 3 ($\Theta = \Theta_3$) increases the number of admissible hypomatchable sets and hence the variety of "admissable" matchings.

In computer–implementations of the primal–dual method it will be convenient to assume the dual variables y_R for $R \in S$ to be <u>nonnegative,</u> hence the updating formulas and the definition of dual feasibility etc. change slightly.

Let us finally state the updating formulas for the Blossom–algorithm based on characterization II.

Here we start with and maintain a dual solution y fulfilling the condition

$$y_i + y_j + \sum_{\{i,j\} \in \delta(R)} y_R \leq c_{ij} \quad \text{for } \{i,j\} \in E$$

$$y_R \geq 0 \quad \text{for } R \in S .$$

As for characterization I we define

$$E(y) = \{\{i,j\} \in E \mid \bar{c}_{ij} = 0\}$$

$$S(y) = \{R \in S \mid y_R > 0\}$$

and again we assume the dual solution to be *strongly dual feasible*, i.e.

$$\overline{S(y)} := S \backslash S(y) \quad \text{is a shrinking family in} \quad G(y) = (V, E(y)) .$$

Now let $\overline{M}$ be a maximum cardinality matching in $G(y) \times \overline{S(y)}$ which has been obtained by the CMP–labeling method and let $\tilde{S} \subset S$ be the set of hypomatchable sets which are shrunk at the end of the CMP–labeling method. Then the sets $\tilde{S}_0(\tilde{S}_I)$ and V_0 (V_I) are defined as above. In addition we define $\bar{V}_0$ $(\bar{V}_I)$ as the set of nodes contained in V_0 (V_I) yet not in any member $R \in \tilde{S}$, i.e. $\bar{V}_0$ $(\bar{V}_I)$ are the original nodes in $G \times \tilde{S}$ which are outer (inner) nodes of a Hungarian tree.

Then we calculate

$$\Theta_1 := \min \ \{\tfrac{1}{2}\bar{c}_{ij} \mid i,j \in V_0 \text{ , not contained in a common}$$
$$\text{hypomatchable set} \ R \in \tilde{S}\}$$

$$\Theta_2 := \min \ \{\bar{c}_{ij} \mid i \in V_0 \text{ , } j \in V \backslash (V_0 \cup V_I)\}$$

$$\Theta_3 := \min \ \{y_R \mid R \in \tilde{S}_I\} .$$

With $\Theta = \min \ \{\Theta_1, \Theta_2, \Theta_3\}$ the improved dual solution $\tilde{y}$ is defined as

$$\tilde{y}_i = \begin{cases} y_i + \Theta & \text{for } i \in \bar{V}_0 \\ y_i - \Theta & \text{for } i \in \bar{V}_I \end{cases}$$

$$\tilde{y}_R = \begin{cases} y_R + \Theta & \text{for } R \in \tilde{S}_0 \\ y_R - \Theta & \text{for } R \in \tilde{S}_I . \end{cases}$$

The updating formula for characterization II does not involve nodes which are contained in hypomatchable sets which are shrunk. Thus this updating formula is somewhat more "economical". Moreover this updating formula is more in the spirit of the philosophy of "shrinking". Throughout the procedure the subgraphs which are shrunk and the dual variables associated with their nodes remain unchanged, i.e. one really works on the "surface graph" $G \times \bar{S}$ only.

If we interpret the dual variables y_R for $R \in \bar{S}$ not as dual variables associated with the hypomatchable set R but with the associated pseudonode then the above two formulas reduce to essentially one formula for updating the dual variables for the nodes / pseudonodes in the surface graph $G \times \bar{S}$.

$$\bar{y}_k = \begin{cases} y_k + \Theta & \text{if } k \text{ is outer (pseudo-) node in } G \times \bar{S} \\ y_k - \Theta & \text{if } k \text{ is an inner (pseudo-) node in } G \times \bar{S} \end{cases} .$$

In both variants of the blossom–algorithm the occurence of $\Theta = \Theta_3$ requires the "expansion" of an inner pseudonode. This is a feature of the weighted matching algorithms which is not needed in CMP–implementations where only <u>outer</u> pseudonodes have be expanded after an augmentation resp. at the end of the whole procedure. This additional task requires the use of more advanced data–structures than those used in the "simpler" CMP–algorithm and causes the main problems with respect to theoretical efficiency. Yet also in practical implementations this expansion–step is the most time–consuming routine of the whole procedure.

Nearly all implementations of Edmonds' blossom algorithm are based on characterization I (cf. Lawler [1976], Weber [1980]). The reason may be the fact that characterization I can also be modified to be valid in the case when optimal non–perfect matchings are required and that Edmonds' original algorithm was based on this characterization. Weber [1981] has analysed the characterization I – algorithm with respect to postoptimality and sensitivity analysis. He ended up with several different cases (and procedures) to be distinguished and rather involved updating formulas. Derigs [1981] showed that for characterization II the updating formulas are again somewhat simpler. (Postoptimality analysis becomes even more elegant if the blossom–algorithm is interpreted as shortest augmenting path method). An analysis of both characterizations with respect to possible implementation strategies can be found in Ball and Derigs [1983].

Each iteration of the blossom algorithm can easily be performed in $O(|V||E|)$ time yielding a total complexity of $O(|V|^2|E|)$ resp. $O(|V|^4)$. Lawler [1976] was the first who presented a $O(|V|^3)$ labeling–technique. Using a new type of priority queue "with variable priority", Galil et al. [1982] could give a $O(|V||E|\log|V|)$ implementation of the blossom–algorithm recently (see also Ball and Derigs [1983]).

From the above it should be evident that applying Edmonds' blossom algorithm to a bipartite graph yields the Hungarian method. And in fact the blossom algorithm was highly influenced by Kuhn's AP–method, more precisely the strategy of how to use the combinatorial min–max relations combined with a dual–updating technique.

13.3. The Primal Method

In this section we review a primal approach for solving MP which has been developed by Cunningham and Marsh [1978] (cf. also March [1978]). Originally the method was formulated to solve the "maximization 1MP" i.e. the problem

$$max \ c'x$$

$$Ax = 1$$

$$x \in \{0,1\}^E$$

with A the node–edge incidence matrix of a graph $G = (V, E)$. Yet we will discuss here the (immediate) translation of this method to the standard "minimization" problem. The relations between this method and Edmonds' blossom–algorithm are the same as between Kuhn's Hungarian method and Balinski and Gomory's approach for the assignment problem. In fact specializing (the minimization variant of) Cunningham and Marsh's procedure yields Balinski and Gomory's method.

While the primal–dual algorithms can be viewed as strategies which starting from a feasible dual solution and a non–perfect matching, together fulfilling complementary slackness, work towards achieving primal feasibility, i.e. a perfect matching, the primal methods maintain a feasible solution, i.e. a perfect matching / assignment, and obtain a feasible solution for the associated dual problem only at termination.

As Edmonds' original version also Cunningham and Marsh work with characterization I. Yet from the results on the primal–dual procedure it should be clear that an "equivalent" procedure based on characterization II can easily be constructed. Since to our knowledge this variant hasn't been discussed so far in literature we will concentrate on this modification here.

As for Balinski and Gomory's method we define for M a matching in G and Q an M–alternating cycle

$$c(Q) := c(Q \backslash M) - c(M \cap Q)$$

to be the *length* of Q . Then the following theorem can be viewed as the combinatorial background of Cunningham and Marsh's procedure

Theorem 13.3.
A perfect matching M in $G(V, E)$ is optimal, i.e. is a min–cost perfect matching, iff G does not contain M–alternating cycles with negative length.

Proof. (see section 10.3.).

Hence Cunningham and Marsh's approach is a so–called "negative cycle algorithm", where given a perfect matching one tries to improve the current solution by an exchange (augmentation) along a negative alternating cycle.

As for the primal AP–method we call a perfect matching M and a dual solution y a *complementary pair* if the reduced cost coefficients of all matching edges are zero i.e. if characterization II is used we require

$$\bar{c}_{ij} = c_{ij} - y_i - y_j - \sum_{(i,j)\in\delta(R)} y_R = 0 \quad \text{for } \{i,j\} \in M .$$

Moreover in the nonbipartite case we require

$$|M \cap \gamma(R)| = \frac{1}{2}(|R| - 1) \quad \text{for } R \in S \text{ with } y_R > 0$$

which is equivalent to $|M \cap \delta(R)| = 1$ for $R \in S$ with $y_R > 0$. Recall that for the Hungarian method we required the initial dual solution to be strongly dual feasible, and the algorithm would maintain this property . Motivated by this definition we require in the primal algorithm the pair M and y to be *strongly complementary*. For that purpose given M a perfect matching and $y \in \mathbb{R}^{V \cup S}$ a complementary pair we define

$$E(y) := \{\{i,j\} \in E \mid \bar{c}_{ij} = 0\} \quad \text{and}$$
$$\overline{S(y)} := \{R \in S \mid y_R > 0\} .$$

As before we call $E(y)$ the set of *admissible* edges and an edge $e \in E$ with $\bar{c}_e \geq 0$ is called *feasible*. Then M and y are called *strongly complementary* iff

$$\overline{S(y)} \quad \text{is a shrinking family in} \quad G(y) = (V, E(y)) .$$

With this definition M induces (uniquely) a matching M_y in $G \times \overline{S(y)}$ and the following essential lemma holds (cf. Lemma 10.6.).

Lemma 13.4.

Let M and y be strongly complementary and let Q_y be an alternating cycle in $G \times \overline{S(y)}$ with respect to M_y then Q_y induces (uniquely) an M–alternating cycle Q in G with the following properties

$$c(Q) = \bar{c}(Q) = \bar{c}(Q_y) .$$

Proof.

(i)   The relation $\bar{c}(Q) = \bar{c}(Q_y)$ follows immediately from the way Q is constructed from Q_y by sucessively expanding pseudonodes contained in Q_y. Thereby always even (alternating) paths are inserted the edges of which have reduced cost zero.

(ii)   The relation $c(Q) = \bar{c}(Q)$ follows by similar argumentation:

For the edges of Q which are contained in $G \times \overline{S(y)}$ no " blossom weights" are incorporated when calculating their reduced cost. When successively expanding pseudonodes and inserting even (alternating) paths the dual weights associated with the blossom just considered cancel out. When having expanded all pseudonodes contained in Q also the "node weights" cancel out as in the bipartite case.

□

Thus we can search for negative cycles in $G \times \overline{S(y)}$ using the reduced cost coefficients. Yet not every (negative) M–alternating cycle Q must be induced by a (negative) M_y–alternating cycle Q_y. Hence when finding no negative M_y–alternating cycle in $G \times \overline{S(y)}$ one has to make sure that no negative M–alternating cycle is "hidden".

An indication for the optimality of a perfect matching is the feasibility of the dual solution y. The primal algorithm can for instance be started with any perfect matching M and a dual solution $y \in \mathbb{R}^{V \cup S}$ fulfilling $y_R = 0$ for all $R \in S$. This is achieved by setting

$$y_i = y_j = \frac{1}{2}c_{ij} \quad \text{for } \{i,j\} \in M .$$

This pair is obviously strongly complementary.

For the discussion of the primal approach it is convenient to introduce the following notation:

For $i \in V$ we define $b(i)$ to be the pseudonode in $G \times \overline{S(y)}$ containing node i where we set $b(i) := i$ if i is contained in the surface graph. For $i \in V$ with

219

$b(i) \neq i$ we define $B(i) := R$ with R the maximal member of $\overline{S(y)}$ containing i . For $i \in V$ with $b(i) = i$ we define $B(i) := \{i\}$.

Now assume M and y a strongly complementary pair and assume $\{j*, k*\} \in E$ with $\bar{c}_{j* k*} < 0$. The primal procedure will now sucessively change the dual solution (and eventually perform one augmentation involving edge $\{j^*, k^*\}$) until finally edge $\{i^*, j^*\}$ becomes feasible while no feasible edge becomes non-feasible again. For that purpose a M_y–alternating tree T in $G(y) \times \overline{S(y)}$ rooted at (pseudo-) node $b(i^*)$ is grown where i^* is defined by $\{b(i^*), b(j^*)\} \in M_y$.

Now two cases may occur

<u>Case 1</u> (Pseudo-) node $b(k^*)$ becomes an outer node of T ,

<u>Case 2</u> T becomes a Hungarian tree in $G(y) \times \overline{S(y)}$.

In the first case we perform the following *mini–dual–change*.

<u>Procedure 13.1.</u>   **Mini–Dual–Change and Augmentation**

<u>Step 1</u> Set $\delta := -\bar{c}_{j*, k*}$

if $b(j^*) = j^*$ goto <u>Step 3</u>

if $y_{B(j*)} \geq \delta$ goto <u>Step 4</u>

else goto <u>Step 2</u>

<u>Step 2</u> Set $y_{B(j*)} := 0$ i.e. $S(y) := S(y) \cup B(j)$.

Extend M_y and redefine $b(j^*)$ and $B(j^*)$,

goto <u>Step 1</u>

<u>Step 3</u> Set $y_{j*} := y_{j*} - \delta$ and goto <u>Step 5</u>

<u>Step 4</u> Set $y_{B(j*)} = y_{B(j*)} - \delta$ and goto <u>Step 5</u>

<u>Step 5</u> Augment M_y along the M_y–alternating cycle induced by the alternating tree and edge $\{b(j^*), b(k^*)\}$.

By this mini–dual–change and augmentation a strongly complementary pair is obtained with the number of admissible edges increased by at least one.

In the second case when T becomes a Hungarian tree in $G(y) \times \overline{S(y)}$ we perform a dual–change similar to the one which is performed during the blossom–algorithm with the additional feature that (pseudo-) node $b(j^*)$ is treated as inner (pseudo-) node of the tree.

<u>Procedure 13.2.</u>   **Dual–change**

Calculate

$\Theta_1 := min \ \{\frac{1}{2}\bar{c}_{ij} \mid b(i) \neq b(j) \ , \ \text{outer (pseudo-) nodes}\}$

$\Theta_2 := min \ \{\bar{c}_{ij} \mid b(i) \ \text{ outer (pseudo-) node, } b(j) \ \text{ not in alternating tree}\}$

$\Theta_3 := min \ \{y_k \mid k \ \text{ inner pseudonode of the alternating tree}\}$

$\Theta := min\{\Theta_1, \Theta_2, \Theta_3\}$

and set

$$y_k := \begin{cases} y_k + \Theta & \text{for } k \text{ outer (pseudo-) node} \\ y_k - \Theta & \text{for } k \text{ inner (pseudo-)node .} \end{cases}$$

By this dual change a strongly complementary pair is obtained where all feasible edges stay feasible. Moreover let Θ be the "value" of the dual change, if $b(j^*) \neq b(k^*)$ and $-\bar{c}_{j^* k^*} < \Theta$, then edge $\{k^*, j^*\}$ becomes feasible. Note that if $b(j^*) = b(k^*)$, all hypomatchable sets $R \in \overline{S(y)}$ containing j^* and k^* have to be expanded before edge $\{j^*, k^*\}$ can become feasible via a (mini-) dual change.

The primal algorithm will achieve dual feasibility, hence optimality, after growing at most $|E|$ trees. As in the blossom algorithm each iteration can easily be performed in at most $O(|V||E|)$ time from which an overall complexity of $O(|V||E|^2)$ resp. $O(|V|^5)$ for the algorithm follows. Obviously the quality of the initial perfect matching has significant influence on the efficiency of the primal algorithm. Here an initial perfect matching can for instance be obtained by applying the CMP–algorithm. Cunningham and Marsh [1979] propose a certain "greedy–like heuristic" where the perfectness of the matching is guaranteed by introducing artificial edges with large costs eventually ("Big M"–start procedure).

In the following we present two variants of this primal method for which the above bound can be improved significantly. <u>Refinement 1:</u> ("most negative rule")

This is in fact the variant which has been proposed by Cunningham and Marsh [1978] originally. As for the Balinski and Gomory's AP–method, it is obvious that if edge $\{k^*, j^*\}$ is chosen in such a way that

$$\bar{c}_{k^* j^*} = min \ \{\bar{c}_{kj^*} \mid k \in N(j^*)\} < 0$$

then every edge in $\delta(j^*)$ has become feasible when edge $\{k^*, j^*\}$ becomes feasible.

With this refinement at most $|V|$ trees have to be grown reducing the amount to $O(|V|^2 |E|)$ steps. Marsh [1979] has shown how this "refinement" can be implemented in $O(|V|^3)$ time using the data–structures proposed by Lawler [1976] for the blossom–algorithm.

<u>Refinement 2:</u> ("dimension expanding approach")

This refinement is motivated by the analog for the bipartite case (cf. section 10.3.). Besides an interesting theoretical property to be discussed lateron, this refinement has some computational advantage, too.

The primal (labeling) algorithm is complicated by the fact that some edges $\{j^*, k^*\}$ contained in hypomatchable sets $R \in S$ with $y_R > 0$ may have negative reduced cost. Let us therefore call a strongly complementary pair M and y a *compatible* pair iff

$$R \in \overline{S(y)} \ \Rightarrow \ \bar{c}_{ij} \geq 0 \quad \text{for all } \{i, j\} \in \gamma(R) \ .$$

Obviously any initial dual solution with $y_R = 0$ for $R \in S$ and $y_i + y_j = c_{ij}$ for $\{i,j\} \in M$ leads to a compatible pair.

The following variant will preserve this property.

Let $i_1, \ldots, i_n$ be any enumeration of V then we define the initial matching and dual solution in the following way:

$$M := \bigcup_{k=1\ldots\frac{n}{2}} \{i_{2k-1}, i_{2k}\}$$

$$y_{i_{2k-1}} := y_{i_{2k}} := \frac{1}{2} c_{i_{2k-1}, i_{2k}} \quad \text{for} \quad k = 1, \ldots, \frac{n}{2}$$

where we introduce artificial edges with sufficiently large cost if necessary ("Big M"–start procedure).

With respect to this enumeration we define

$$V_k := \{i_1, i_2, \ldots, i_{2k}\} \quad \text{for} \quad k = 1, \ldots, \frac{n}{2},$$

$$G_k = (V_k, \gamma(V_k) \cap E) \quad \text{for} \quad k = 1, \ldots, \frac{n}{2},$$

$$S_k = \{R \in S \mid R \subset V_k\} \quad \text{for} \quad k = 1, \ldots, \frac{n}{2}.$$

Then $M^1 = \{i_1, i_2\}$ is an optimal perfect matching in G_1 with (y_{i_1}, y_{i_2}) an optimal dual solution.

Now assume an optimal perfect matching M^k in G_k and $y^k \in \mathbb{R}^{V_k \cup S_k}$ an associated dual solution. Then a *compatible* pair $M^{(k+1)}$ and $y^{(k+1)}$ in G_{k+1} is obtained in the following way:

$$M^{k+1} := M^k \cup \{i_{2k+1}, i_{2k+2}\}$$

$$y_j^{k+1} := y_j^k \quad \text{for} \quad j \in V_k$$

$$y_{i_{2k+1}}^{k+1} := min \left\{ c_{j, i_{2k+1}} - y_j - \sum_{j \in R, R \in S_k} y_R^k \mid j \in V_k \right\}$$

$$y_{i_{2k+2}}^{k+1} := c_{i_{2k+1}, i_{2k+2}} - y_{2k+1}^{k+1}$$

$$y_R^{k+1} = \begin{cases} y_R^k & \text{for } R \in S_k \\ 0 & \text{for } R \in S_{k+1} \setminus S_k. \end{cases}$$

An optimal solution (perfect matching) in G_{k+1} can now be obtained by one iteration of the primal approach. Compared to the other variants these tree growing procedures are somewhat "simpler" since the "root node" is always an original node and all edges contained in blossoms which are shrunk are feasible.

Although the worst-case complexity is the same as for refinement 1, in practice the number of dual-changes and tree-growing steps might be smaller since the maximal possible size of the alternating trees is smaller and the number of possible blossoms to be shrunk as well. We will discuss this refinement from a theoretical point of view lateron again.

13.4. The Shortest Augmenting Path Approach

In this section we introduce the *shortest augmenting path concept* (SAP) for solving MP. The idea of such an approach was already mentioned by Glover [1967]. Since no efficient method for solving "shortest alternating path problems" in nonbipartite graphs was known, Brown [1977] proposed to solve these subproblems by enumerative (branch and bound) techniques. Gondran and Minoux [1980] explicitly formulated the existence of a good SAP-algorithm in nonbipartite graphs as an "open problem".

An efficient SAP-labeling technique was first given in Derigs [1981] with a FORTRAN-code contained in Burkard and Derigs [1980]. In the following we will first present the basic theory and then explicitly formulate this "Dijkstra-like" labeling procedure.

In section 10.5. we have defined a matching M in a bipartite graph to be extreme iff it does not allow any negative M-alternating cycle and we have shown that a perfect matching is optimal iff it is extreme. Now the SAP-approach is again based on the following theorem:

Theorem 13.5.

Let M be an extreme matching with $s, t \in U(M)$ and let P be a shortest M–alternating path connecting s and t, then $M := M \oplus P$ is extreme again.

Proof. (similar to the bipartite case, cf. Theorem 10.13.).

Then the SAP–method will start with an extreme matching M and successively determine "shortest augmenting paths" and augment the matching. As in the bipartite case we call this a *phase* of the procedure. Thus at most $\frac{|V|}{2}$ phase–subproblems — determination of a shortest augmenting path — have to be solved. Here we can distinguish different variants of subproblems — the node s (and t) may be prespecified or the shortest augmenting path connecting any possible pair of M–unsaturated nodes may be determined.

As in the bipartite case we call a matching M *k–optimal* iff $|M| = k$ and $c(M) \leq c(M')$ for all $M' \in \mathcal{M}$ with $|M'| = k$. Obviously any k–optimal matching is extreme.

For $W \subseteq V$ we define the *W–restricted problem* as follows. Let $G' = (W, \gamma(W))$ and $\mathcal{M}'_p$ the set of perfect matchings in G' then the problem is

$$min \ \{c(M) \mid M \in \mathcal{M}'_p\}$$

which can also be written in the following way

$$min \sum_{\{i,j\} \in \gamma(W)} c_{ij} \cdot x_{ij}$$

$$\sum_{\{i,j\} \in \delta(i) \cap \gamma(W)} x_{ij} = 1 \quad \text{for} \ \ i \in W$$

$$x_{ij} \in \{0,1\} \quad \text{for} \ \{i,j\} \in \gamma(W) .$$

Introducing the appropriate set of blossom–inequalities or cut–inequalities, respectively, we can write the above $(0,1)$–program as linear program again and W–optimal matchings can be characterized via the existence of a complementary dual solution.

Now assume an extreme matching M in G with $V(M)$ the set of nonexposed nodes with respect to M. Then M is obviously an optimal solution for the $V(M)$–restricted problem. Define S' as the set of hypomatchable sets in $G' = (V(M), \gamma(V(M)))$ then exists $y' \in \mathbb{R}^{V(M) \cup S'}$ which is feasible for the dual program associated with the matching problem on G'. Let $\bar{c}_{ij}$ denote as usual the reduced cost of an edge $\{i, j\} \in \gamma(V(M))$ with respect to y' and $E(y') := \{\{i, j\} \in \gamma(V(M)) \mid \bar{c}_{ij} = 0\}$. Then the following properties hold:

(i) $M \subseteq E(y')$

(ii) $\overline{S'(y')} = \{R \in S' \mid y'_R > 0\}$ is a shrinking family

$$\text{in } G'(y') = (V(M), E(y'))\} .$$

If $V(M) \neq V$ i.e. M is not an optimal solution for the given problem, this dual solution y' can easily be extended to a vector $y \in R^{V \cup S}$ which is feasible for the dual program of the original matching problem on G by setting

$$y_i := \begin{cases} y_i & \text{if } i \in V(M) \\ -L & \text{if } i \in V \backslash V(M) \end{cases}$$

$$y_R := \begin{cases} y_R & \text{if } R \in S' \\ 0 & \text{if } R \in S \backslash S' . \end{cases}$$

where L is a sufficiently large number (Note that for the original matching problem and the $V(M)$–restricted problem the same characterization has to be chosen).

Then $\overline{S(y)} := \{R \in S \mid y_R > 0\}$ is a shrinking family in $G(y) = (V, E(y))$ where $E(y) = \{\{i, j\} \in E \mid \bar{c}_{ij} = 0\}$ and $\bar{c}_{ij}$ is the reduced cost with respect to y, now.

Thus we can formulate the following lemma:

Lemma 13.6.

A matching M in $G = (V, E)$ is extreme iff exists $y \in \mathbb{R}^{V \cup \mathcal{S}}$ with the properties

(i) $y_R \geq 0$ *for* $R \in \mathcal{S}$

(ii) $\bar{c}_{ij} \geq 0$ *for* $\{i, j\} \in E$

(iii) $M \subseteq E(y)$

(iv) $\overline{\mathcal{S}(y)}$ *is a shrinking family in* $G(y)$

(v) $R \in \overline{\mathcal{S}(y)} \Rightarrow |M \cap \delta(R)| = 1$ *and* $|M \cap \gamma(R)| = q_R$

(vi) $i, j \in V \backslash V(M) \Rightarrow y_i = y_j (=: y_0)$.

A pair (M, y) fulfilling the above properties is called a *strongly compatible pair*.

Note that the calculation of $\bar{c}_{ij}$ depends on what characterization has been chosen. Yet for any strongly compatible pair (M, y) with respect to characterization I we can easily construct a $\hat{y}$ such that $(M, \hat{y})$ is strongly compatible with respect to characterization II, and vice versa.

Again we define for M-alternating paths $P = (V(P), E(P))$

$$c(P) := c(E(P) \backslash M) - c(E(P) \cap M)$$

the *length* of P. For the following discussion we assume an extreme(non-perfect) matching M and a dual vector y such that (M, y) is a strongly compatible pair. Then we denote by M_y the matching in $G \times \overline{\mathcal{S}(y)}$ which is induced by M.

Lemma 13.7.

Let P be an M-augmenting path connecting the M-unsaturated nodes s and t which is induced by an M_y-augmenting path in $G \times \overline{\mathcal{S}(y)}$. Then

$$c(P) = \bar{c}(P) + 2y_0$$

holds.

Proof. We give the proof here for characterization II only. The proof for characterization I follows by similar arguments.

$$\bar{c}(P) = \sum_{\{i,j\}\in E(P)\setminus M} \bar{c}_{ij} - \sum_{\{i,j\}\in E(P)\cap M} \bar{c}_{ij}$$

$$= \sum_{\{i,j\}\in E(P)\setminus M} c_{ij} - \sum_{\{i,j\}\in E(P)\cap M} c_{ij} - y_s - y_t$$

$$+ \sum_{R\in \mathcal{S}} (|(E(P)\cap M)\cap \delta(R)| - |(E(P)\setminus M)\cap \delta(R)|) \cdot y_R$$

Since all hypomatchable sets $R \in \mathcal{S}$ with $y_R \neq 0$ are shrunk we have

$$|(E(P)\cap M)\cap \delta(R)| = |(E(P)\setminus M)\cap \delta(R)| = 1 \quad \text{for all} \quad R \quad \text{with} \quad y_R \neq 0$$

which implies

$$\bar{c}(P) = c(P) - y_s - y_t = c(P) - 2y_0 \ .$$

$\square$

This relation motivates the use of reduced cost coefficients $\bar{c}_{ij}$ instead of the actual costs c_{ij} when looking for the shortest M–augmenting path P which is induced by an M_y–alternating path in $G \times \overline{S(y)}$. With respect to the reduced costs the length of a path becomes a "simple" sum since $\bar{c}_{ij} = 0$ for $\{i,j\} \in M$ i.e.

$$\bar{c}(P) = \sum_{\{i,j\}\in E(P)} \bar{c}_{ij} \ .$$

Thus again only with respect to $\bar{c}_{ij}$ the notion "length" of P is really adequate.

Lemma 13.8.

Let P be an M–augmenting path which is induced by an M_y–augmenting path in $G \times \overline{S(y)}$ and let R be a maximal element in $\overline{S(y)}$. By P_R we denote the partial path of P which is contained in R . Then

$$\bar{c}(P_R) = 0$$

holds.

Proof. Follows immediately from the fact that $\overline{S(y)}$ is a shrinking family in $G(y)$ and hence for every $R \in S$ the odd cycle spanning $G[R] \times S[R]$ is contained in $E(y)$. Since the partial path of P contained in R is part of that odd cycle we have $E(P_R) \subset E(y)$.

$\square$

Thus if the shortest augmenting path P with respect to M in G is induced by an M_y–augmenting path in $G \times \overline{S(y)}$ then we can find P by looking for the shortest M_y– augmenting path in $G \times \overline{S(y)}$ with respect to the reduced cost values $\bar{c}_{ij}$. The SAP–method will exactly follow this advice and start to find the shortest M_y–augmenting path in $G \times \overline{S(y)}$.

Yet the shortest M–augmenting path need not be induced by a M_y–augmenting path. The shortest augmenting path algorithm must be able to detect whether this is the case and it must overcome this situation. For that purpose the method maintains "control variables" indicating the possibility of such "hidden paths", which cannot be found when working in $G \times \overline{S(y)}$ only. In that case certain pseudonodes are expanded to enable their detection.

Lemma 13.9.

Let P be an M–augmenting path which is <u>not</u> induced by an M_y–augmenting path in $G \times \overline{S(y)}$ but which is induced by an augmenting path in $G \times (\overline{S(y)} \backslash \{R\})$ where R is a maximal member of $\overline{S(y)}$ then

$$c(P) = \begin{cases} \bar{c}(P) + y_R + 2y_0 & \text{for characterization I} \\ \bar{c}(P) + 2y_R + 2y_0 & \text{for characterization II .} \end{cases}$$

Proof. We give the prove here again for characterization II only. The proof for characterization I follows by similar arguments.

Since $|(E(P) \cap M) \cap \delta(R)| = 0$ and $|(E(P) \backslash M) \cap \delta(R)| = 2$ the result follows immediately by applying the formula given in the proof of lemma 13.7., thus

$$\bar{c}(P) = c(P) + 2y_R + 2y_0 \ .$$

☐

Let us call such a path P to be "hidden by shrinking R". Thus as long as working in $G \times \overline{S(y)}$ can guarantee to produce an augmenting path P the "reduced–cost–length" $\bar{c}(P)$ of which does not exceed the "reduced–cost–length" of any augmenting path which is "hidden" by any maximal member $R \in \overline{S(y)}$ by more than y_R resp. $2y_R$, the shortest augmenting path which is found in $G \times \overline{S(y)}$ will induce a shortest augmenting path in G , too.

Motivated by the above lemmata and given a strongly compatible pair (M, y) the SAP–method is the combination of three basic modules:

- a *Dijkstra–like labeling method* for constructing the shortest alternating path tree *(forest)* in $G \times \overline{S(y)}$ with respect to the reduced cost coefficients,

- an *expansion–subroutine* which expands a maximal member $R \in \overline{S(y)}$ and extends the alternating path tree from $G \times \overline{S(y)}$ onto $G \times \overline{S(y)} \backslash \{R\}$,

- a *dual–update procedure* which given a shortest M–alternating path P constructs a dual solution $\hat{y}$ such that $(M \oplus P, \hat{y})$ is strongly compatible, again.

After developing the basic ideas of the shortest–augmenting path approach for the nonbipartite case we will now state the method more formally by giving a labeling–technique for solving the phase–subproblem, i.e. determining a shortest augmenting path with respect to an extreme matching M provided a strongly compatible dual solution y is at hand.

We assume (M, y) a strongly compatible pair with respect to characterization II. From the presentation above it should be clear that a related labeling procedure based on characterization I can easily be formulated. Yet we feel that especially for the SAP–concept characterization II is more adequate. W.l.o.g. we assume $y_0 = 0$, i.e. $y_i = 0$ for $i \in U(M)$. It can be shown, that the dual variables for unmatched nodes remain unchanged after the augmentation via P and the following dual update. Then if we start with $M = \emptyset$ and $y \equiv 0$, this assumption holds throughout the whole process.

Throughout the procedure we work on a surface graph $\overline{G} := G \times \overline{S}$ where $\overline{S} \subset S$ is a shrinking family. Moreover every $R \in \overline{S}$ is a blossom with respect to the matching M with which the procedure is started. Initially we have $\overline{S} = \overline{S(y)}$, yet during the course of the procedure "new" hypomatchable sets $R \in S$ may be detected and shrunk, while maximal members $R \in \overline{S(y)}$, resp. the associated pseudonodes, may be expanded again. In $G \times \overline{S}$ we distinguish two kinds of nodes — original nodes $i \in V$ and pseudonodes corresponding to maximal members $R \in \overline{S}$; those pseudonodes will be denoted by R , too. Thus the dual variable y_R can be interpreted as "pseudonode (dual) variable". By $\overline{V}$ we will denote the set of nodes in $\overline{G}$, original nodes as well as pseudonodes. Thus " $i_0 \in \overline{V}$ " may be used for $i_0 \in V$ as well as for $i_0 = R$, a maximal member of $\overline{S}$. For every node $w \in \overline{V}$ we define by $\overline{N}(w)$ the set of adjacent (pseudo-) nodes in $G \times \overline{S}$.

As for the primal method we define for all $i \in V$

$$b(i) := \begin{cases} i & \text{if } i \in \overline{V} \\ R & \text{if } i \in R , R \text{ maximal member in } \overline{S} \end{cases}$$

$$B(i) := \begin{cases} \{i\} & \text{if } i \in \overline{V} \\ b(i) & \text{if } b(i) \neq i . \end{cases}$$

Formally $b(i)$ is a (pseudo-) node in $\overline{G}$, while $B(i)$ is a subset of nodes in G .

Let $\overline{M}$ be the matching in $G \times \overline{S}$ which is induced by the matching M . Due to the fact that (M, y) is a strongly compatible pair $U(M) \subseteq \overline{V}$ holds, i.e. all M–unsaturated nodes are contained as original nodes in $G \times \overline{S(y)}$. Note that $\overline{S}$ and hence $\overline{G}$ and $\overline{M}$, too may change throughout the procedure. Thus a node $s \in U(M)$ may be shrunk in the course of the labeling process.

As in the bipartite case we can distinguish two SAP–versions

- the *tree–version*, where a shortest M–alternating path starting at a prespecified node $s \in U(M)$ is constructed by building up an alternating tree in $G \times \overline{S}$ rooted at s ,

- the *forest–version*, where a shortest M–alternating path connecting any two nodes $s, t \in U(M)$ is constructed by building up an alternating forest in $G \times \overline{S}$ the trees of which are rooted at nodes $s \in U(M)$.

In the following we will describe the tree–version in detail and thereafter we will outline the modification for the forest–version.

Let us call an $\overline{M}$–alternating path even (odd) depending on its number of edges in $\overline{G}$. Then each node $w \in \overline{V}$ may receive the following (temporary) labels:

$d_w^- :=$ current lower bound for the length of the shortest odd $\overline{M}$–alternating path connecting w and the "root" s ,

$d_w^+ :=$ current lower bound for the length of the shortest even $\overline{M}$–alternating path connecting w and the "root" s

where all "distances" are measured with respect to the reduced cost–coefficients. Nodes $w \in \overline{V}$ with finite d_w^- (d_w^+)–label are called "$-$" labeled resp. "$+$" labeled. Note that all inner nodes of the alternating tree are "$-$" labeled while all outer nodes are "$+$" labeled.

For $w \in \overline{V}$ and w "$-$" labeled we set

$p(w) :=$ predecessor of w on the alternating path defining d_w^- .

Here the p–label is always pointing to an original node $i \in V$ and then the associated (pseudo-) node v of the surface graph is given by $v := b(p(w))$. Now if $w = R$ with R a maximal member of $\overline{S}$ we also have to store the node $l \in R$ which defines d_R^- . This is done by setting $q(R) = l$. For convenience we define q–labels for original nodes in the surface graph, too.

The following figure may illustrate the use of the different labels.

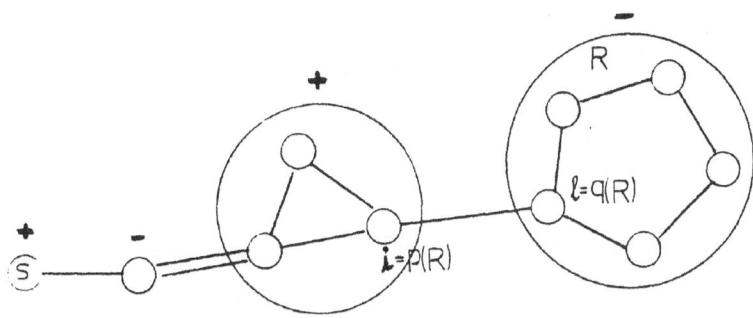

Figure 13.2. Alternating tree in $G \times \overline{S}$

During the course of the procedure these labels are "updated" by "scanning" outer nodes $j \in \overline{V}$ of the $\overline{M}$-alternating tree. Now $j \in \overline{V}$ may stand for a maximal member $R \in \overline{S}$. In that case all the original nodes contained in R resp. their neighbourhood has to be examined. Formally this basic procedure can be stated as follows

---

SCAN (j)

  For all $k \in \overline{N}(j)$
    for all $\ell \in B(k)$
      for all $i \in B(j)$
        set $d_k^- := min\{d_k^- , d_j^+ + \bar{c}_{il}\}$ and
        $p(k) := i$ and $q(k) := l$ if d_k^-
      has changed.

With $s \in U(M)$, the prespecified root–node, the labeling procedure is started with the following.

233

Set $d_j^+ := d_j^- := \infty$ for all $j \in \bar{V}$.

Make s an outer node of the $\bar{M}$-alternating tree and set $d_s^+ := 0$.

SCAN (s) .

The progress of the procedure is controlled by the following central "branching-routine".

<u>CONTROL</u>

Determine

$\delta_1 := min \ \{d_i^- \mid i \in U(\bar{M}) \backslash \{s\}\}$

$\delta_2 := min \ \{d_i^- \mid i \in \bar{V} \backslash U(\bar{M}) \ , \ i \ \text{not a tree-node}\}$

$\delta_3 := min \ \{\frac{1}{2}(d_i^- + d_i^+) \mid i \in \bar{V} \ , \ \text{outer node}\}$

$\delta_4 := min \ \{d_R^- + y_R \mid R \in \bar{V} \ , \ \text{inner pseudonode}\}$

If all sets are void: STOP, G does not contain any perfect matching.

Set $\delta := min \ \{\delta_1, \delta_2, \delta_3\}$ and let $i_0 \in \bar{V}$ be the node in $\bar{G}$

defining δ .

If $\delta_4 < \infty$, let R_0 be the pseudonode defining δ_4 .

(Note that i_0 may be a pseudonode, too.)

If $\delta_4 < \delta \quad \rightarrow \quad$ EXPAND (R_0)

if $\delta = \delta_1 \quad \rightarrow \quad$ AUGMENT (i_0)

if $\delta = \delta_2 \quad \rightarrow \quad$ GROW (i_0)

if $\delta = \delta_3 \quad \rightarrow \quad$ SHRINK (i_0) .

Throughout the procedure the value δ will be a lower bound for the shortest M-augmenting path starting at node s .

The value $\delta_4 = min \ \{d_R^- + y_R \mid R \ \text{inner pseudonode} \ \}$ is what we have

234

called a "control variable". Here d_R^- is the length of the M–alternating path connecting the nodes in R with the root–node s if the path–length is measured by the reduced cost coefficients. Thus let P be a M– augmenting path which is hidden by R , i.e. which can be detected after expanding the (maximal) hypomatchable set R only. Then due to lemma 13.9. (and the assumption $y_0 = 0$)

$$c(P) = \bar{c}(P) + 2y_R \ .$$

Now $\bar{c}(P) \geq d_R^-$ hence

$$c(P) \geq d_R^- + 2y_R \quad \text{holds} .$$

As long as $\delta \leq \delta_4$, the procedure will continue to grow the alternating tree in the current surface graph $G \times \overline{S}$, to find the shortest $\overline{M}$–augmenting path starting at $b(s)$. If a $\overline{M}$–augmenting path is detected in the next step, this path will induce a shortest M–augmenting path, too.

On the other hand if $\delta_4 < \delta - y_R$, a "hidden" M– augmenting path P with $c(P) < \delta$ may exist. During the procedure we use the case $\delta_4 < \delta$ already as a "signal" to modify the surface graph by expanding the inner pseudonode R_0 defining δ_4 .

In the following we state and interpret the subroutines which are called by CONTROL:

__GROW (i_0)__

    Let $j_0 \in \overline{V}$ be defined by $\{i_0, j_0\} \in \overline{M}$.

    Make i_0 an inner node and j_0 an outer node of the $\overline{M}$–alternating tree.

    Set $d_{j_0}^+ := \delta$.

    SCAN (j_0) .

    Return to CONTROL.

In the GROW–routine the minimal tentative d^-–label which is obtained for the non–tree node i_0 , is fixed, i.e. becomes permanent. This is a common step borrowed from Dijkstra's label–setting algorithm for determining (simple) shortest paths in graphs with nonnegative edge weights. Since the shortest alternating path of even length connecting s and j_0 must contain the matching edge $\{i_0, j_0\}$ and consequently the shortest alternating path of odd length connecting s and i_0 , the label $d_{j_0}^+$ can be fixed, too. Since the reduced cost of the matching edge "connecting" i_0 and j_0 is zero, we get $d_{j_0}^+ = d_{i_0}^- (= \delta)$.

We now discuss the implications of $\delta = \delta_3$.

Let $j_0 := p(i_o)$, then introducing edge $\{i_0, b(j_0)\}$ to the $\bar{M}$–alternating tree creates a unique odd cycle $K := K(i_0, p(i_0))$, i.e. a blossom with respect to $\bar{M}$ in $\bar{G}$, which induces a blossom $R := R(K) := \bigcup_{j \in K} B(j)$ with respect to M in G . With these definitions we can formulate the SHRINK–operation:

<u>SHRINK (i_0)</u>

For all $i \in K$, set

$$
y_i := \begin{cases} y_i + (\delta - d_i^+) & \text{for } i \text{ outer (pseudo-) node} \\ y_i + (d_i^- - \delta) & \text{for } i \text{ inner (pseudo-) node.} \end{cases}
$$

Set $\overline{S} := \overline{S} \cup \{R\}$ and make R an outer node of the $\bar{M}$–alternating tree.
Set $y_R := 0$ and $d_R^+ := \delta$.
For all $j \in K$, j formerly an inner (pseudo-) node of the tree, SCAN (j) .
Return to CONTROL.

Obviously $\delta = \delta_3$ is possible in nonbipartite graphs only. When introducing the edge $\{i_0, b(j_0)\}$ in addition to the outer nodes of the tree also the inner nodes contained in K are "reachable" from s by an <u>even</u> $\bar{M}$– alternating path, using edges in the tree and edge $\{i_0, b(j_0)\}$ only.

Now assume a node v which <u>before</u> the SHRINK–operation is not contained in K but is adjacent to a node in K . Thus exists edge $\{t, k\} \in E$ with $t \in B(v)$ and $b(k) \in K$. This situation is illustrated by the following figure:

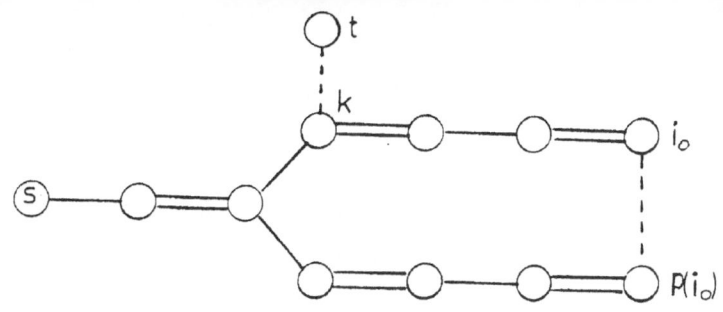

Figure 13.3.

Now we can show that the operation performed by SHRINK does not change the length $\bar{L}$ of the alternating path from s to v ending with edge $\{k, t\}$. More precisely we can show that the altered labels give the correct path lengths.

Before the application of the SHRINK–procedure we had

$$\bar{L} = \begin{cases} d_{b(k)}{}^+ + \bar{c}_{kt} & \text{for } b(k) \text{ outer node} \\ d_{i_0}^- + (d_{i_0}^+ - d_{b(k)}^-) + \bar{c}_{kt} & \text{for } b(k) \text{ inner node.} \end{cases}$$

After the application of the SHRINK–procedure the reduced cost $\bar{c}_{kt}$ has changed to $\tilde{c}_{kt}$ say, and we get

$$\tilde{L} = d_R^+ + \tilde{c}_{kt} = \delta + \tilde{c}_{kt} = \begin{cases} \delta + (\bar{c}_{kt} - (\delta - d_{b(k)}^+)) & \text{if } b(k) \text{ outer node} \\ \delta + (\bar{c}_{kt} - (d_{b(k)}^- - \delta)) & \text{if } b(k) \text{ inner node.} \end{cases}$$

Using the fact that $2\delta = d_{i_0}^- + d_{i_0}^+$, we get the desired identity $\tilde{L} = \bar{L}$. We have already discussed the implications of the case $\delta_4 < \delta$ and we will now describe the subroutine EXPAND (R_0) by which the maximal member $R_0 \in \bar{S}$ will be "expanded".

EXPAND (R_0)

    Set $\delta := \delta_4$.

    Set $y_{R_0} := 0$ and $\bar{S} := \bar{S} \backslash \{R_0\}$.

    Extend matching $\bar{M}$ to the odd cycle spanning $G[R_0] \times S[R_0]$

237

(i.e. $\bar{G}$ and $\bar{M}$ change.)

The odd cycle is canonically partitioned into an odd alternating path and an even alternating path. Make the even alternating path part of the tree and set d_j^+ resp. d_j^- to δ for the outer resp. inner (pseudo-)nodes of this path and set $p(j)$ and $q(j)$ properly.

SCAN (j) for outer nodes of this even alternating path.

For the non–tree nodes j of the odd alternating path

set $d_j^- := min \ \{d_{b(i)}^+ + \bar{c}_{ik} \mid k \in B(j)$ and

$$b(i) \in \bar{N}(j) \text{ outer (pseudo-) nodes } \}$$

and set $p(j)$ and $q(j)$ properly.

Return to CONTROL.

Arguments and calculations similar to the case $\delta = \delta_3$ show that the operations performed by the EXPAND produce correct labels.

Let $\bar{c}_{ij}$ be the reduced costs before and $\tilde{c}_{ij}$ be the reduced costs after the EXPAND–operation. Thus the reduced costs of all edges in $\delta(R_0)$ have been increased by y_{R_0}. Let k be a node which is adjacent to R_0 before the EXPAND–operation and let P be the M–alternating path connecting k and s, which was hidden in the old surface graph due to the fact that R_0 was shrunk.

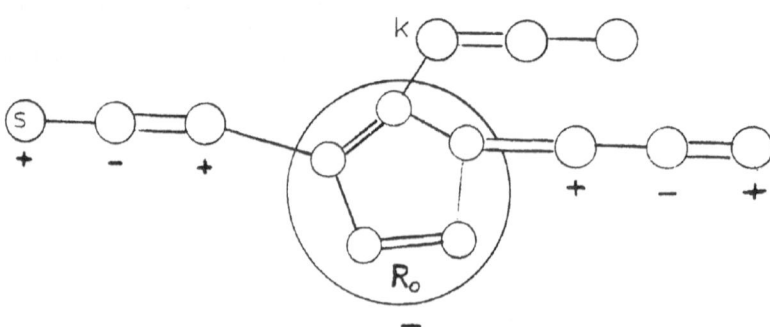

Figure 13.4. Augmenting path P connecting s and t before expansion of R_0

After "expanding R_0" let j be the node in the new surface graph which is adjacent to k in the alternating path P .

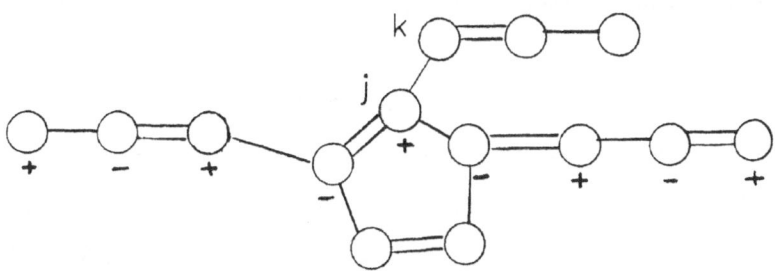

Figure 13.5. Augmenting path P after expansion of R_0

Then we have

$$d_j^+ = \delta = d_{R_0}^- + y_{R_0} \quad \text{and}$$

$$\tilde{c}_{jk} = \bar{c}_{jk} + y_{R_0} \ .$$

Thus $\tilde{c}(P) = \bar{c}(P) + 2y_{R_0}$ which is the correct path length according to lemma 13.9. .

Finally, we discuss the case $\delta = \delta_1$. Here a shortest augmenting path has been detected and the current phase terminates.

AUGMENT (i_0)

Exchange the matching and nonmatching edges on the alternating path P connecting s and i_0 .

Perform the following DUAL–UPDATE.

For all $i \in \bar{V}$

Set $\quad y_i := \begin{cases} y_i + (\delta - d_i^+) & \text{if } i \text{ outer (pseudo-) node} \\ y_i + (d_i^- - \delta) & \text{if } i \text{ inner (pseudo-) node} \\ y_i & \text{if } i \text{ non–tree (pseudo-) node} \end{cases}$

STOP.

The above interpretation should have shown already that the labeling procedure is working correctly, i.e. producing a shortest augmenting path. What remains to show is that after the DUAL–UPDATE in AUGMENT a strongly compatible pair is constructed again. In fact showing this last property also proves the optimality of P.

At the beginning of the labeling procedure and throughout the procedure we maintain the matching M and a dual solution y such that (M, y) is strongly compatible. In steps SHRINK and EXPAND the dual solution and the surface graph are "locally" changed by a formula similar to the DUAL–UPDATE–formula. Thereby an *alternate dual solution* is constructed which together with M forms a strongly compatible pair again.

In general it is easy to show, that at any stage of the labeling procedure after calculating the δ–value in CONTROL a DUAL–UPDATE using that specific δ–value and the current tree–labeling would yield a dual solution y which is strongly compatible with M and which has the additional property that all edges contained in the current tree or in the odd–cycles spanning hypomatchable sets $R \in \overline{S(y)}$ have zero reduced cost–values.

When performing the DUAL–UPDATE procedure in the AUGMENT–routine a dual solution is obtained which is strongly compatible with both, the old and the new matching in G .

For the *forest version* of the SAP–method we have to modify the INITIALISATION only, where we make each node $s \in U(M)$ the outer root–node of a $\overline{M}$–alternating tree with $d_s^+ := 0$. In CONTROL we replace the calculation of δ_1 and δ_3 by the following formulas

$$\delta_1 := min \ \{\frac{1}{2}(d_i^- + d_i^+) \mid i \in \overline{V} , \text{ outer node with } b(p(i)) \text{ in different tree as } i\}$$

$$\delta_3 := min \ \{\frac{1}{2}(d_i + d_i^+) \mid i \in \overline{V} , \text{ outer node, with } b(p(i)) \text{ in same tree as } i\} .$$

Again, we have detected a shortest M–alternating path if $\delta = \delta_1$ and a blossom with respect to $\overline{M}$ if $\delta = \delta_3$.

Starting with $M = \emptyset$ and $y \equiv 0$ the optimal matching is constructed after $\frac{|V|}{2}$ phases, i.e. applications of the SAP–labeling method. One application of the tree–version as well as of the forest version can easily be performed in $O(|V||E|)$ time, which leads to an algorithm of complexity $O(|V^2||E|)$. Using the data-structures proposed by Lawler [1976] for the blossom algorithm, the complexity can be reduced to $O(|V|^3)$. In Ball and Derigs [1983], we discuss several strategies for implementing matching algorithms based on the SAP concept, one with an $O(|V|^3)$ time bound and one with an $O(|V||E| \, log\,|V|)$ time bound. Both of these implementations have storage requirements that are linear in $|V|$ and $|E|$ (while Lawler's data–structure leads to storage requirements which are quadratic in $|V|$). The concept that is crucial to achieve the $O(|V||E| \, log\,|V|)$ time bound, is a *splittable priority queue*, first introduced by Galil et al. [1982].

Note that the SAP–labeling method will work correctly too in the case $y_0 \neq 0$, and even if $y_i \neq y_j$ for $i,j \in U(M)$. The only requirement is the strong compatibility of M and y. Yet if $y_i = y_j$ for $i,j \in U(M)$, this property will be preserved after the DUAL–UPDATE. Special refinements of the general SAP–labeling method are obtained with special initial pairs (M,y).

If we apply the forest–version to a strongly compatible pair (M,y) with $y_i = y_j$ for $i,j \in U(M)$ a sequence of k–optimal is constructed. A common initialisation is given with $M = \emptyset$ and $y \equiv 0$. Yet from a computational point of view, special start–routines which produce strongly compatible pairs (M,y) with "large" $|M|$ are favorable.

13.5. Near Equivalence of Matching Algorithms

In this section we show that the primal–dual blosssom algorithm and a certain (efficient) refinement of the primal algorithm can be viewed as special implementations of the shortest augmenting path concept.

Another look at the blossom–algorithm shows that all matchings which are produced in the course of the procedure are extreme. This is easily seen since throughout the procedure a strongly compatible pair (M, y) is at hand. As in the bipartite case, starting with $y \equiv 0$ a sequence of k–optimal solutions will be constructed.

As in the bipartite case, let us analyse the labeling process of the blossom–method more precisely by looking at the sequence of steps during one phase. Here a phase of the blossom–algorithm is the sequence of steps between two augmentations without a change in $\overline{S(y)}$, i.e without a change of the surface graph on which the CMP–labeling procedure is performed.

Again we discuss the implementation based on characterization II only. For that purpose let $(M^{(k)}, y^{(k)})$ be the strongly compatible pair at the beginning of a phase with $M^{(k)}$ the matching which is induced by a maximum cardinality matching $\overline{M}$ in $G(y^{(k)}) \times \overline{S(y^{(k)})}$. Obviously we can rewrite the formulas for calculating the Θ–value of the blossom–algorithm in the following way using the notation of the SAP–approach:

$\Theta_1 := min \; \{\bar{c}_{ij} \mid b(i) \;$ outer (pseudo-) node, $b(i) \;$ non–tree (pseudo-) node$\}$

$\Theta_2 := min \; \{\bar{c}_{ij}/2 \mid b(i) \neq b(j) \; ,$ outer (pseudo-) nodes$\}$

$\Theta_3 := min \; \{y_R \mid R \;$ inner pseudonode .$\}$

Let $\Theta_1^*, \ldots, \Theta_r^*$ be the sequence of Θ– values which have defined the first r dual changes of the current phase and let $D := \sum_{i=1}^r \Theta_i^*$. For any (pseudo-) node q in $\overline{G}$ we denote by $L(q)$ the number of dual updates performed during the present phase before (pseudo-) node q became a node of one of the alternating trees. Then

$$d_q := \sum_{i=1}^{L(q)} \Theta_i^*$$

is the current value of D at the time node q, first became a node of one of the alternating trees. Now let $\bar{c}_{ij}^{(k)}$ denote the reduced cost coefficients with respect to $y^{(k)}$, i.e. at the beginning of the phase, then we can rewrite the above formulas as follows:

$$\Theta_1 := min \ \{\bar{c}_{ij}^{(k)} - (D - d_{b(i)}) \mid b(i) \ \text{outer (pseudo-)node} ,$$
$$b(j) \ \text{non-tree (pseudo-)node}\}$$

$$= min \ \{d_{b(i)} + \bar{c}_{ij}^{(k)} \mid b(i) \ \text{outer (pseudo-node)} ,$$
$$b(j) \ \text{non-tree (pseudo-)node}\} - D$$

$$\Theta_2 := min \ \{(\bar{c}_{ij}^{(k)} - (D - d_{b(i)}) - (D - d_{b(j)}))/2 \mid b(i) \neq b(j) \ \text{outer}$$
$$\text{(pseudo-)nodes}\}$$

$$= min \ \{((d_{b(j)} + \bar{c}_{ij}^{(k)}) + d_{b(i)})/2 \mid b(i) \neq b(j) \ \text{outer (pseudo-)nodes}\} - D$$

$$= \min_{\substack{b(i) \ \text{outer} \\ \text{(pseudo-)node}}} \{(d_{b(i)} + min\{d_{b(j)} + \bar{c}_{ij}^{(k)} \mid$$
$$b(j) \neq b(i) \ \text{outer (pseudo-)node} \})/2\} - D$$

$$\Theta_3 := min \ \{y_R^{(k)} - (D - d_R) \mid R \ \text{inner pseudonode}\}$$
$$= min \ \{d_R + y_R^{(k)} \mid R \ \text{inner pseudonode}\} - D \ .$$

Now defining $\delta_i := \Theta_i + D$ we get the same minimization formulas as in the CONTROL–step of the SAP–forest version (yet in only a different order) where we do not distinguish between d^+- and d^--labels.

The significance of computing the δ_i's rather than the Θ_i's is that the formulas for the δ_i's do not explixitely require the current dual variables y_j but rather the $y_j^{(k)}$'s and the d_j's only. Thus it is not necessary to update all dual variables after each GROW–step since the d_j's remain constant for (pseudo-) nodes j which are already in the alternating forest.

Hence the SAP–labeling technique can also be interpreted as a more economical implementation of the blossom–algorithm where a sequence of dual–changes is comprised into a general dual–update after the shortest augmenting path has been found and some "local" dual–updates when the surface–graph changes.

Whenever a pseudonode is expanded the d–values of the (pseudo-) nodes which were formerly not contained in the surface–graph are set to δ , which is the current D–value of the blossom–algorithm. Whenever a newly formed blossom is shrunk the DUAL–UPDATE for the (pseudo-) nodes which were formerly in the surface graph and become shrunk now can be performed immediately since the dual variables of these (pseudo-) nodes are not altered by any further step of the blossom–algorithm, and this is exactly done in the local dual–update of the SAP–labeling technique. An analogous relation holds for the blossom–algorithm and the SAP–method based on characterization I.

This discussion and interpretation shows as in the bipartite case (the even more than) near equivalence of the SAP–approach and the primal–dual blossom–algorithm. For a further analsyis of various strategies for implementing the primal–dual matching algorithm based on both characterizations and their interpretation within the SAP–concept we refer to Ball and Derigs [1983].

Let us now analyse the relationship between the primal algorithm of Cunningham and Marsh and the SAP–approach. We will first discuss refinement 2 — the dimension expanding approach. In this variant a sequence $\{M^k\}_{k=1,\dots,n/2}$ of optimal perfect matchings in certain subgraphs $G_k = (V_k, E_k)$ is constructed together with a dual solution $y^k \in \mathbb{R}^{V_k \cup \mathcal{S}_k}$. Extending y^k to $\hat{y}^k \in \mathbb{R}^{V \cup \mathcal{S}}$ by setting

$$\hat{y}_i := 0 \quad \text{for} \quad i \in V \setminus V_k$$

$$\hat{y}_R := 0 \quad \text{for} \quad R \in \mathcal{S} \setminus \mathcal{S}_k$$

yields a strongly compatible pair $(M^k, \hat{y}^k)$. Hence one iteration of the primal method is essentially the determination of a shortest M^k-augmenting path connecting $i_{2(k+1)-1}$ and $i_{2(k+1)}$ combined with a sequence of dual-updates.

Finally we want to outline another interesting relation between the primal approach and the SAP-concept. In analogy to related LP-techniques we will call this refinement the "Big-M-approach". For that purpose we write the matching problem as a linear program again

$$min \; c'x$$

$$Ax = 1$$

$$Bx \geq 1$$

$$x \geq 0$$

with A the node–edge incidence matrix of G and B the matrix representing the cut–inequalities.

Introducing artificial variables we can formulate this problem equivalently as follows

$$min \; c'x + M'z$$

$$Ax + Ez = 1$$

$$Bx \quad \geq 1$$

$$x, \; z \geq 0 \; .$$

Here $M' := (M, \ldots, M)$, with M sufficiently large. For this problem, which is not a 1- matching problem anymore, $x \equiv 0$ and $z \equiv 1$ is a feasible solution. Yet a simple "manipulation" will transform this problem back into a 1MP again. For that purpose we introduce a second set of artificial variables and formulate the following LP:

$$min\ c'x + M'z + O'\tilde{z}$$

$$
\begin{aligned}
Ax\ +\ Ez\qquad\qquad &=\ 1\\
Ez\ +\ D\tilde{z}\ &=\ 1\\
Bx\qquad\qquad\qquad &\geq\ 1\\
x,\ z,\ \tilde{z} \geq 0\,.
\end{aligned}
$$

Here D is any $(0,1)$–matrix with the property that each column contains exactly two nonzero elements and each row contains exactly one nonzero element.

Now the above problem is equivalent to a 1MP on a graph $G' = (V', E')$ obtained from G by introducing $|V|$ "artificial" nodes and $3\,|V|\,/2$ " artificial" edges. Here the edges in $E'\backslash E$ correspond to the artificial variables z and $\tilde{z}$, respectively, and form two matchings M_z and $M_{\tilde{z}}$, with M_z a perfect matching in G' ; while the nodes in $V'\backslash V$ are pairwisely matched by $M_{\tilde{z}}$. The following figure shows the graph G' .

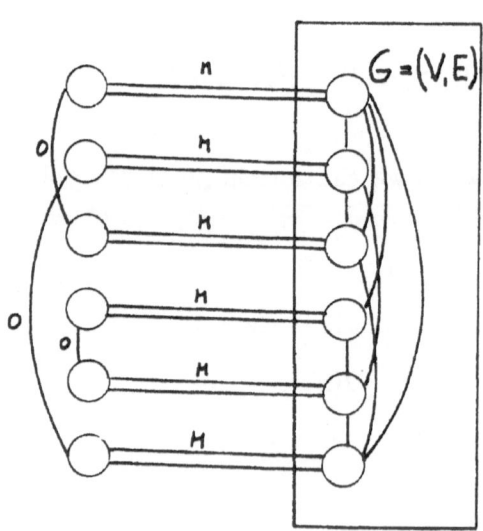

Figure 13.6. Graph $G' = (V', E')$

Note that the set of hypomatchable sets for G' is the same as for G thus the above LP is the correct linear program associated with the 1MP on G'. Since no edge from M_z will be contained in any optimal perfect matching M' in G' we get

$$M' = M_{\bar{z}} \cup M^*$$

with M^* an optimal perfect matching in G. Thus the original 1MP on G can be solved by solving the 1MP on G' instead. Now define

$$y_i := 0 \quad \text{for } i \in V$$

$$y_i := M \quad \text{for } i \in V' \backslash V$$

$$y_R := 0 \quad \text{for } R \in S.$$

Then (M_z, y) is a strongly complementary pair which is compatible as well.

Applying Cunningham and Marsh's primal method to this problem starting with the pair (M_z, y) will produce the optimal matching $M' = M_{\bar{z}} \cup M^*$, hence an optimal perfect matching in G in $n' := |V| / 2$ steps.

In the course of the procedure a sequence M^k, $k = 0, \ldots, n'$, of perfect matchings in G' is constructed with $M^0 = M_z$ and $M^{n'} = M_{\bar{z}} \cup M^*$. In general M^k can be partioned into three sets

$$M^k = M_{\bar{z}}^k \cup M_z^k \cup M_G^k \qquad \text{with}$$

$M_{\bar{z}}^k \subseteq M_{\bar{z}}$, $M_z^k \subseteq M_z$ and M_G^k a matching in G.

Then an iteration of the primal algorithm can be interpreted as constructing the shortest M_G^k-augmenting path connecting two prespecified nodes.

After demonstrating the near–equivalence of the three MP–approaches we will finally report some computational results. There are only very few contributions in the literature which describe and report numerical tests with "matching–codes" Cunningham and Marsh [1978] compared implementations of their primal method and the blossom algorithm (tree–version) with both implementations using the same standards (data–structures etc.). Their tests showed the blossom–code to be clearly superior to the primal–code in average performance, taking from 2/3 to 1/6 of the time. To their opinion the main reason for this result lies in the fact that the blossom algorithm can grow precisely $|V|/2$ trees, whereas the primal algorithm can grow as many as $|V| - 1$ trees. Yet what is even more crucial, the primal algorithm performs many more grow–steps, shrink-steps and expand-steps than the blossom algorithm does, since a perfect matching is maintained.

In Derigs and Kazakidis [1979] several versions of the SAP–approach and the blossom–algorithm are compared, again using the same standards (data structures etc.). Here the SAP–codes showed to be superior to the blossom–codes, with the forest–versions outperforming the tree–versions. A direct comparison between the different codes showed the Derigs and Kazakidis – codes to be superior to the Cunningham and Marsh – codes which is essentially due to the more efficient data–structures used in our implementations, (cf. Derigs [1983]).

13.6. Epilogue: The Matching Simplex Method

The search for nice simplex–methods for the nonbipartite matching problem — i.e. a simplex method encountering combinatorially attractive bases only – is motivated by two facts:

(i)   the intractability of the general simplex method for the nonbipartite matching problem and

(ii)   the existence of a nice polynomial simplex method for the bipartite 1–matching problem.

While the LP–formulation of network flow problems is of the same size as the integer programming formulation, the size of the LP–formulation of MP is exponential with respect to the order of the graph (number of nodes). Thus a "brute force" application of the simplex method to MP is computationally intractable. Since throughout the procedure an exponential number of slack–variables associated with the blossom–inequalities or cut–inequalitites, respectively, are basic variables, exponentially many of these inequalities have to be considered when determining the basic variable which is to leave the basis during a pivot step. Thus tractability is the dominant purpose for seeking a nice MP–simplex method.

On the other hand it could be shown that for AP certain nice simplex–methods turn out to be even polynomial algorithms. Thus MP in nonbipartite graphs is obviously one of the next candidates for such polynomial simplex methods. Yet the research on this field is quite young and the results obtained so far are not as promissing as one might have expected (and hoped for).

The results which we outline here are due to Edmonds and Koch [1981] contained in an unpublished working paper (cf. also Koch [1979] for the generalization to the bMP). Edmonds and Koch have discovered classes of combinatorially attractive bases which are convenient from a computational point of view, so–called "nice" MP–bases with the property that for every MP

- an "initial" basis can easily be constructed,

- given a non–optimal nice basis there always exists a (polynomial) simplex pivot to another nice basis,

- there always exists an optimal basis which is "nice".

Yet "cycling" could not be prevented by a pivoting rule which preserves the property that only nice bases are encountered. Thus the finiteness of these MP–simplex methods, hence the existence of a nice MP–simplex method at all, is still an open question.

Let us consider the linear system defining the matching polytope (using the cut–inequalities)

$$Ax = 1$$

$$Bx \geq 1$$

$$x \geq 0 .$$

Here we may assume that A is of full row rank since it is the node–edge incidence matrix of a (complete) nonbipartite graph $G = (V, E)$. Let the columns of A and B be indexed by E and the rows of A and B be indexed by V and S, respectively. Then we call a (column–) basis of the above system a *MP–basis*.

A *pseudotree* is a connected graph containing no even polygon and exactly one odd polygon. A graph the components of which are pseudotrees is called a *pseudoforest*. Pseudoforests have the following useful property

Lemma 13.10.

Let $G' = (W, F)$ be a pseudoforest. Given $d \in \mathbb{R}^W$ and $c \in \mathbb{R}^F$, then exists unique $x \in \mathbb{R}^F$ such that

$$x(\delta_{G'}(v)) = d_v \quad \text{for every } v \in W$$

and exists a unique $y \in \mathbb{R}^W$ such that

$$y_i + y_j = c_{ij} \quad \text{for } \{i, j\} \in F .$$

For $F \subseteq E$ we define $G_F := (V, F)$ the subgraph induced by F. Let $\mathcal{R} \subseteq \mathcal{S}$ then $(F, \mathcal{R})$ is called a *bundle* iff

(i) $\overline{\mathcal{R}} := \mathcal{S} \backslash \mathcal{R}$ is a nested family with $\overline{\mathcal{R}}_V := \overline{\mathcal{R}} \cup \{V\}$.

(ii) $G_F[S] \times \overline{\mathcal{R}}_V[S]$ is a pseudoforest for all $S \in \overline{\mathcal{R}}_V$.

Edmonds and Koch [1983] have shown the following basic properties of bundles:

Theorem 13.11.

For every bundle $(F, \mathcal{R})$ the system of equations

$$x_e = 0 \qquad \text{for } e \notin F$$

$$x(\delta(v)) = 1 \quad \text{for } v \in V$$

$$x(\delta(S)) = 1 \quad \text{for } S \in \overline{\mathcal{R}}$$

has a unique solution and for every $c \in \mathbb{R}^E$ the system of equations:

$$y_R = 0 \quad \text{for every } R \in \mathcal{R}$$

$$y_i + y_j + \sum_{\{i,j\} \in \delta(R)} y_R = c_{ij} \quad \text{for } \{i,j\} \in F$$

has a unique solution.

Thus every bundle is a MP–basis (more precisely, every bundle induces a MP–basis). Moreover if $(F, \mathcal{R})$ is a feasible bundle, i.e. it induces a feasible MP–basis, then the associated basis is integer valued.

A bundle $(F, \mathcal{R})$ is called a *connected bundle* iff

$$G_F[S] \times \overline{\mathcal{R}}_V[S] \quad \text{is a pseudotree for every } S \in \overline{\mathcal{R}}_V$$

and $(F, \mathcal{R})$ is called a *polygonal–bundle* iff

$$G_F[S] \times \overline{\mathcal{R}}_V[S] \quad \text{is an odd polygon for every } S \in \overline{\mathcal{R}}_V.$$

Analysing the final output of the blossom–algorithm, Edmonds and Koch could show the following essential result.

Theorem 13.12.

For any objective function $c : E \to \mathbb{R}$, the matching–LP has an optimal basis which is induced by a polygonal bundle.

An initial feasible polygonal bundle is constructed by the following simple "start procedure":

Let $v_1, v_2, \ldots, v_n$ be any enumeration of V. Then introduce $n/2$ artificial edges

$$E_1 := \{\{v_{2i}, v_{2i-1}\} \mid i = 1, \ldots, n/2\}$$

and another $n/2$ artificial edges

$$E_2 := \{\{v_{2i+1}, v_{2i-1}\} \mid i = 1, \ldots, n/2 - 1\} \cup \{v_1, v_5\} .$$

Then $(E_1 \cup E_2, S)$ is a feasible polygonal bundle which forms a pseudotree with $\{v_1, v_3, v_5\}$ defining the odd polygon. The associated basic solution is given by

$$x_{ij} = 0 \quad \text{for} \quad \{i, j\} \in E \cup E_2 \quad \text{and}$$
$$x_{ij} = 1 \quad \text{for} \quad \{i, j\} \in E_1 .$$

Let $(F, \mathcal{R})$ be a polygonal bundle then $(F^*, \mathcal{R}^*)$ with $F^* = F \cup \{e\}, e \notin F$ and $\mathcal{R}^* = \mathcal{R}$ or $F^* = F$ and $\mathcal{R}^* = \mathcal{R} \cup \{S\}, S \notin \mathcal{R}$ is called a *near–bundle*. Given a near–bundle $(F^*, \mathcal{R}^*)$ define $\overline{\mathcal{R}}^* := S \backslash \mathcal{R}^*$ and $\overline{\mathcal{R}}_V^* := \overline{R}^* \cup V$. Then exactly one out of the following two cases occurs :

(i) For a single $S \in \overline{\mathcal{R}}_V^*$, exactly one component of $G_{F^*}[S] \times \overline{\mathcal{R}}_V^*[S]$ contains an even polygon.

(ii) For a single $S \in \overline{\mathcal{R}}_V^*$, exactly one component of $G_{F^*}[S] \times \overline{\mathcal{R}}_V^*[S]$ contains two edge disjoint odd polygons joined by a single path — a so–called *dumbell*.

Pivoting between two (polygonal) bundles involves going from a bundle to a near–bundle and from the near–bundle to a bundle again.

Describing six different possible types of simplex–pivots Edmonds and Koch could verify the following result.

Theorem 13.13.

Given a non–optimal feasible polygonal bundle there always exists a simplex pivot to another feasible polygonal bundle.

The optimality of a polygonal bundle can be checked in polynomial time due to the fact that the sets associated with the non–basic slack–variables form a nested family $\overline{\mathcal{R}}$ and hence $|\overline{\mathcal{R}}| \leq |V| - 1$ holds.

As mentioned already, Edmonds and Koch did not succeed in finding a pivot–rule which avoids cycling without forcing to switch to a non–nice MP–basis eventually. The existence of such a pivot–rule would imply, that given a non–optimal nice MP–basis B_0 , there always exists a finite sequence $B_1, B_2, \ldots, B_n$ of nice bases with B_n being an optimal MP–basis, such that B_{i+1} is obtained from B_i by a simplex pivot. Even the existence of such a "nice simplex–path" is still an open question. Koch [1979] reports that a computer program implementing a nice simplex method which is not proven to terminate, did perform successfully on a variety of 1MP's.

PART VI

THE *b*–MATCHING PROBLEM

In this section we discuss the last cornerstone problem within the class of general matching problems — *the b-matching problem (bMP)*. In fact this problem can also be designated as the key problem of the entire class, since every other problem is either a special case of bMP or can be reduced to bMP by a *polynomial* transformation. Thus giving an efficient algorithm for bMP makes the whole class a well–solved class of combinatorial optimization.

It is evident that, due to the fact that bMP can be reduced to 1MP, all 1MP-approaches can be generalized to bMP and the near equivalence of these methods for bMP would follow immediately from the results on 1MP.

Blossom algorithms — i.e. primal–dual–methods for bMP occur in Edmonds, Johnson and Lockhart [1969] and Pulleyblank [1973]. Later, Havel [1975] has "cleaned up" the Blossom III–code of Edmonds et al. [1969].

These algorithms are not "good" in the sense of Edmonds since the computational bound depends linearly on the "size" of b , i.e. is not polynomial on the input size of the problem. Marsh [1979] combined the blossom–algorithm for bMP and the scaling methodology of Edmonds and Karp [1972] — originally developed for network flow problems — to obtain a theoretical improvement in that the computational bound depends logarithmically on the b's which makes the approach a good algorithm.

Koch [1979] has investigated the possibility of constructing a "nice" *b*–matching simplex method. The results are as in the 1MP–case: Koch succeeded in describing classes of nice bMP-bases with the property that each bMP has an optimal nice basis with an initial nice basis easily constructible. He could also describe (combinatorially natured) pivot-operations leading from a nice basis to another nice basis. Yet the existence of a "finite nice simplex path" is still an open problem.

To our knowledge the primal 1MP–algorithm of Cunningham and Marsh has never been extended to bMP.

We will develop here the shortest augmenting path method for bMP. From the results on 1MP one could also describe this method as a (more economical) implementation of the primal–dual blossom algorithm. Yet we argue again, that the shortest augmenting path concept — i.e. this greedy–like appraoch — is the combinatorial backbone of matching-algorithms and we will show again that the scaling method can be viewed as a special implementation of the shortest augmenting path method.

As in the HTP–case the scaling method was never viewed as a serious candidate for the most efficient b–matching algorithm and the result was more of theoretical interest. Under this point of view the importance of the scaling technique is more or less lost since the development of the ellipsoid-method (cf. Khachian [1979]) and the establishment of its impact for combinatorial optimization in general by Grötschel et al. [1981] and especially the development of a polynomial cutting plane technique for solving bMP by Padberg and Rao [1982]. For computational experience with this approach we refer to Holland [1983]. Yet although these results are (even more than) promising — when playing the (dirty) trick of replacing the polynomial ellipsoid method by the non-polynomial simplex method — combinatorial methods are the only candidates which are both practical and polynomial.

Chapter 14. Basic Structures and Operations

Let $G = (V, E)$ be a graph and $(b_v \mid v \in V)$ an integer vector of so-called *node constraints* or *degree requirements*. Then a mapping $x : E \to \mathbb{N}_0$ resp. vector $x \in \mathbb{N}_0^E$ is called a perfect b-matching on G if

$$x(\delta(i)) = b_i \quad \text{for } i \in V.$$

Given a cost function $c : E \to \mathbb{R}_+$ the perfect b-matching problem, which we denote by P_b is given by

$$min \quad c'x \qquad\qquad (P_b)$$
$$x(\delta(i)) = b_i \quad \text{for } i \in V$$
$$x \geq 0, \quad \text{integer valued.}$$

In fact any "vector" $x \in \mathbb{N}_0^E$ satisfying $x(\delta(i)) \leq b_i$ for $i \in V$ is called a b-matching in G. These definitions generalize the respective definitions for the 1-matching case.

Given a b-matching x and $i \in V$ we define

$$d_i(x) := b_i - x(\delta(i))$$

the *deficiency* of x at i and a node $i \in V$ with $d_i(x) > 0$ is called x–deficient. Then $d(G, b, x) := \sum_{i \in V} d_i(x)$ is called the *deficiency* of x and it is a measure of "how far" x is away from being a perfect b-matching. In analogy to the 1-matching case we call

$$x(E) = 1/2(b(V) - d(G, b, x))$$

the *cardinality* of x.

Note that for the same graph $G = (V, E)$ different vectors, b and b', of node-constraints may be given. In that case we will write P_b and $P_{b'}$ and we distinguish between b-matchings and b'–matchings.

The standard b-matching problem as formulated above generalizes the perfect 1-matching problem. Marsh [1979] treats the related problem of finding a (not necessarily perfect) b-matching of maximum weight i.e.

$$max \quad c'x$$

$$x(\delta(i)) \leq b_i \quad \text{for } i \in V$$

$$x \geq 0, \quad \text{integer valued,}$$

while Pulleyblank allows b-matchings which are required to fulfill $x(\delta(i)) = b_i$ for only a subset $V^= \subseteq V$. Yet as in the 1-matching case it is easy to see that all these problems are equivalent. Again we have chosen the minimization problem as "standard bMP", since it fits into our general scheme.

14.1. Combinatorial Analysis

We start our discussion of the b-matching problem by introducing some combinatorial structures occuring in connection with b-matchings and we will state some fundamental combinatorial properties which are essential for the development of b-matching algorithms. Yet we will not explicitly formulate a labeling algorithm for solving the cardinality b-matching problem since the cardinality b-matching problem has found no special interest in the literature. Yet it is obvious that such an algorithm is essential for a "graphical" implementation of the primal–dual concept and thus such a method is inherent in Pulleyblank's implementation of the blossom–algorithm for instance. Again we feel that a separate treatment of the cardinality b-matching problem with the perspective of developing efficient data–structures would be a first step in the direction of more efficient bMP–codes.

Note that from a combinatorial point of view "b-matchings in G" are equivalent to "degree constrained subgraphs" in a related *multigraph* G' and thus the maximum cardinality b-matching can be treated as a "maximum cardinality degree–constrained subgraph problem" (cf. Urquhart [1967]). An augmenting path method for solving the latter problem can also be found in Berge [1972].

The following theorem which characterizes graphs which allow perfect b–matchings, is due to Tutte [1954]. This theorem was later proved in an algorithmic way by Pulleyblank [1973]. Recently Anstee [1983] gave an algorithm which either finds a perfect b-matching or shows that there does not exist any perfect b-matching in $O(|V|^3)$ time, thereby establishing Tutte's condition.

For any $W \subseteq V$ we define

$$C(W) = \{S \subseteq V \setminus W \mid G[S] \text{ is a component of } G[V \setminus W]\}$$
$$C_0(W) = \{S \in C(W) \mid |S| = 1\}$$
$$C_1(W) = \{S \in C(W) \mid |S| > 1, b(S) \text{ odd }\}.$$

Theorem 14.1. *(Tutte [1954])*
$G = (V, E)$ has a perfect b–matching iff for each $W \subseteq V$

$$b(W) \geq |C_1(W)| + b(\cup C_0(W)).$$

The above theorem is actually an extension of Tutte's 1–factor theorem (cf. section 12.1.) and Pulleyblank [1973] could prove the following lemma which is the counterpart of Berge's lemma (cf. corollary 12.2.) and characterizes the "b–matching number" of G , i.e. the "cardinality" of a maximum b–matching.

Lemma 14.2. *(Pulleyblank [1973])*
For any graph $G = (V, E)$ and $b \in \mathbb{N}^V$

$$max\{x(E) \mid x(\delta(i)) \leq b_i, x \text{ integer valued }\}$$

$$= (b(V) + \min_{W \subseteq V}\{b(W) - |C_1(W)| - b(\cup C_0(W))\})/2 .$$

Let $G' = (V', E')$ be a subgraph of G and x a b-matching in G with the property

$$x(\delta_{G'}(i_0)) = b_{i_0} - 1 \text{ for } i_0 \in V'$$

$$x(\delta_{G'}(i)) = b_i \text{ for } i \in V' \setminus \{i_0\}$$

then x is called a *near-perfect b-matching of G' deficient at i_0* .

We call G' *b-critical (hypomatchable)* iff for <u>every</u> $i_0 \in V'$ there exists a near-perfect b-matching of G' deficient at i_0 . A set $V' \subseteq V$ is called b-critical if $G[V']$ is b-critical. This definition implies that $b(V')$ is odd since with x a near-perfect b-matching in $G' = (V', E')$ we get $2x(E') = b(V') - 1$ hence $b(V') \equiv 1 \bmod 2$. This definition extends the definition of hymatchable sets in the 1-matching case to bMP. Due to the fact that for bMP the "hypomatchability" of a subgraph is depending not only on the graphical structure but also on the degree-constraints we prefer the notion "b-critical" furtheron.

Using Tutte's theorem the following lemma can be shown.

Lemma 14.3.   *(Pulleyblank [1973])*
A connected graph $G = (V, E)$ is b-critical iff $b(V)$ is odd, $|V| \neq 1$ and for every $\emptyset \neq W \subseteq V$

$$b(W) \geq b(\cup C_0(W)) + |C_1(W)| + 1.$$

Recall that a pseudotree is a connected (sub-) graph containing no even polygon and exactly one odd polygon. In section 13.6. we have mentioned that given a pseudotree $B = (V', E')$ and degree-constraints b_i for $i \in V'$, there exists a unique mapping $x' : E \to \mathbb{R}^{E'}$ such that

$$x'(\delta_{G'}(i)) = b_i \text{ for every } i \in V'.$$

Moreover if x' is nonnegative, then x' is integer valued. Thus it follows immediately that special pseudotrees $G' = (V', E')$ with $b(V')$ odd are b-critical.

Now a pseudotree $B = (V', E')$ with odd polygon $P = (V(P), E(P))$ is called a *blossom* with respect to a b–matching x in G if

(i) x is a near–perfect matching of G' which is deficient at a node $u \in V(P)$,

(ii) $x_e \geq 1$ for each $e \in E' \setminus E(P)$,

(iii) $x_e \geq 1$ for each $e \in E(P)$ which is the first edge in the even length path in P from a node $i \in V(P) \setminus \{u\}$ to u .

The following figure gives a sample blossom. Note that in connection with b–matchings edges e with $x_e \geq 1$ are drawn by double lines.

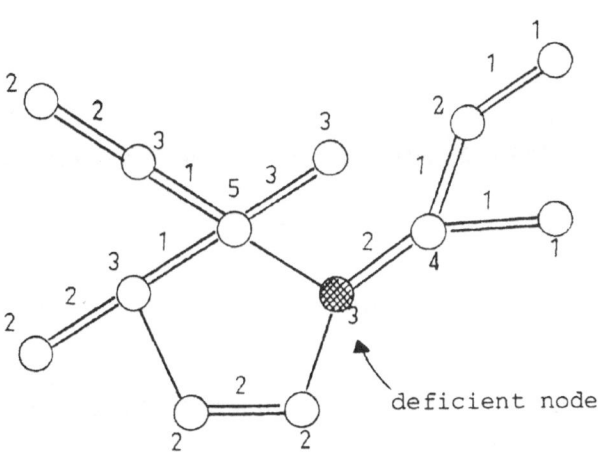

Figure 14.1. Sample blossom

It can be shown that any pseudotree B which is a blossom with respect to a b–matching x in G is b–critical.

For any $i_0 \in V'$ the near perfect matching $x^{(i_0)}$ of B which is deficient at i_0 can be constructed by the following simple procedure given the near–perfect matching x which is deficient at $u \in V'$:

If $i_0 = u$, then $x^{(i_0)} = x$. Otherwise, let $P(u, i_0)$ be the even length path in B from u to i_0 having the smallest possible number of edges. Then define

261

$$x_e^{(i_0)} := \begin{cases} x_e + 1, & \text{if } e \text{ is an odd edge of } P(u, i_0) \\ x_e - 1, & \text{if } e \text{ is an even edge of } P(u, i_0) \\ x_e, & \text{otherwise.} \end{cases}$$

Note that the near–perfect matching $x^{(i_0)}$ can also be constructed from scratch without using the b-matching x . Moreover if $i_0 \in V(P)$ then G' is again a blossom with respect to $x^{(i_0)}$.

Let us call a pseudotree $B = (V', E')$ a b-critical pseudotree if exists a b-matching x such that B is a blossom with respect to x , i.e. B is b-critical.

Now Pulleyblank [1973] has shown the following theorem.

Theorem 14.4.

A graph $G' = (V', E')$ is b-critical iff exists a nested family $\mathcal{A} \subseteq 2^{V'}$ such that

(i) $V' \in \mathcal{A}$ and

(ii) for every $A \in \mathcal{A}$, $G'[A] \times A[A]$ is spanned by a b-critical pseudotree $B[A]$.

A nested family $\mathcal{A}$ of G' with the property (ii) is called a *shrinking family*.

As in the 1–matching case, we define

$$G \times \mathcal{A} := (\ldots (G \times A_1) \times \ldots \times A_q)$$

with $\{A_1, \ldots, A_q\}$ the set of maximal members in $\mathcal{A}$ and these maximal b-critical sets in $\mathcal{A}$ are called *pseudonodes* in $G \times \mathcal{A}$. In $G \times \mathcal{A}$ we then define $b_{A_i} := 1$ for $i = 1, \ldots, q$.

Given a graph $G = (V, E)$ and $b \in \mathbb{N}^V$ we denote by $S(b)$ the set of b-critical sets in G . These results generalize structures and theorems from 1MP–theory. For $b \equiv 1$, b-critical pseudotrees reduce to odd–cycles and theorem 12.4. becomes a special case of theorem 14.4. . Moreover the following relation holds obviously for bMP, too.

Lemma 14.5.

Any b–matching x_A in $G \times A$ can be extended to a b–matching x in G with

$$d(G, b, x) = d(G \times A, b, x_A).$$

14.2. The b–Matching Polytope

With A the node–edge incidence matrix of G , the b–matching problem can be formulated as

$$min \ c'x$$

$$Ax = b$$

$$x \in \mathbb{Z}^E.$$

Again we consider the linear programming relaxation of bMP and we call the system

$$\mathcal{F}(A, b) = \{x \in \mathbb{R}^E \mid Ax = b, x \geq 0\}$$

the *fractional b–matching polytope* associated with G and b . As already seen for the special case $b \equiv 1$, $\mathcal{F}(A, b)$ may have fractional extreme points — so–called fractional b–matchings – if $b \not\equiv 0$ mod 2 . Extending Balinski's theorem 13.1. one can show that every fractional b–matching x is "half integer" i.e. $2x \in \mathbb{Z}^E$ holds.

From a theoretical and computational point of view it is an interesting fact that an optimal fractional b–matching can be constructed by means of solving a Hitchcock–transportation problem.

For that purpose define $G' = (V_s, V_t, E')$ as the symmetric bipartite graph with V_s and V_t two copies of V . For $i \in V$ we denote by i_s (i_t) the copy of i in V_s (V_t) . Then E' is defined by the following relation

$$\{i, j\} \in E \quad \Leftrightarrow \quad \{i_s, j_t\}, \{j_s, i_t\} \in E'$$

and we define
$$b'_{i_s} := b'_{i_t} := b_i \quad \text{for } i \in V$$
$$c'_{i_s j_t} := c'_{j_s i_t} := c_{ij} \quad \text{for } \{i,j\} \in E.$$

With these definitions the associated HTP is:

$$min \sum_{\{i_s, j_t\} \in E'} c'_{i_s j_t} \cdot x'_{i_s}$$

$$\sum_{\{i_s, k\} \in E'} x'_{i_s k} = b'_{i_s} \text{ for } i_s \in V_t$$

$$\sum_{\{k, j_t\} \in E'} x'_{k j_t} = b'_{j_t} \text{ for } j_t \in V_s$$

$$x'_{i_s j_t} \geq 0 \text{ for } \{i_s, j_t\} \in E'.$$

If the above problem has no feasible solution, then $\mathcal{F}(A, b) = \emptyset$ and hence G does not allow b–matchings, too. Otherwise let x^* be an optimal solution for the above HTP, then an optimal fractional b-matching in G is given by

$$x_{ij} := 1/2(x^*_{i_s j_t} + x^*_{j_s i_t}) \text{ for } \{i,j\} \in E.$$

Recall that if $b \equiv 0 \mod 2$ an optimal fractional b– matching x in G which is integer valued, i.e. an optimal perfect b–matching, can be obtained by solving the HTP on G' with $b'_{i_s} := b'_{i_t} := b_i/2$ and then setting

$$x_{ij} := x^*_{i_s j_t} + x^*_{j_s i_t} \quad \text{for } \{i,j\} \in E,$$

where x^* is again an optimal HTP–solution.

Yet for the following we are interested in the linear description of the convex hull of b–matchings, i.e. b-matching polytope, for $b \not\equiv 0 \mod 2$. We will first study the monotone b–matching polytope which we denote by $P^{\leq}(G, b)$.

Lemma 14.6.

Let $x : E \to \mathbb{Z}^E$ be a mapping satisfying

$$x(\delta(i)) \le b_i \quad \text{for } i \in V,$$

i.e. x is a (not necessarily perfect) b–matching, then for any $W \subseteq V$ such that $b(W)$ is odd

$$x(\gamma(W)) \le (b(W) - 1)/2$$

holds.

Proof. Since $\sum_{i \in W} x(\delta(i)) = 2 \cdot x(\gamma(W)) + x(\delta(W)) \le b(W)$, it follows that $2x(\gamma(W)) \le b(W) - x(\delta(W)) \le b(W)$. Now $x(\gamma(W))$ is integer valued and $b(W)$ is odd. Hence $2x(\delta(W)) \le b(W) - 1$.

$\square$

Let us call a set $W \subseteq V$ with $b(W)$ odd an *odd–set* and define $q_W := (b(W) - 1)/2$. Then the inequalities

$$x(\gamma(W)) \le q_W \text{ for } W \subseteq V, \text{ odd}$$

are valid for the (monotone) b–matching polytope. Edmonds [1965] was the first to show that these "blossom–inequalities" are sufficient to describe the (monotone) b–matching polytope.

Pulleyblank [1973] characterized the facets of $P^{\le}(G, b)$ i.e. the unique minimal subset of blossom inequalities which serves to describe the (monotone) b–matching polytope. Let us call a cutnode i of a graph a *strong cutnode* if $b_i = 1$, then the result is as follows:

Theorem 14.7. *(Pulleyblank [1973])*

The blossom inequality

$$x(\gamma(W)) \leq q_W$$

is a __facet__ of $P^{\leq}(G, b)$ for $W \subseteq V$, odd, $|W| \geq 3$ iff $G[W]$ is b-critical and does not contain a strong cutnode.

From Pulleyblank's theorem follows that the following system of (in)equalities is a linear description of the (perfect) b–matching polytope $P^{=}(G, b)$:

$$x(\delta(i)) = b_i \quad \text{for } i \in V$$

$$x(\gamma(W)) \leq q_W \quad \text{for } W \in \mathcal{S}(b)$$

$$x \geq 0.$$

As for the 1-matching case we call the above system the *blossom– characterization* of $P^{=}(G, b)$. Hence the perfect b–matching problem has the following dual (b–matching) problem:

$$max \sum_{i \in V} b_i \cdot y_i - \sum_{W \in \mathcal{S}(b)} q_W \cdot y_W \quad \text{subject to} \qquad \text{(DbMPI)}$$

$$y_i + y_j - \sum_{\{i,j\} \in \gamma(W)} y_W \leq c_{ij} \quad \text{for } \{i, j\} \in E$$

$$y_W \geq 0 \quad \text{for } W \in \mathcal{S}(b)$$

and optimal perfect b–matchings x^* can be characterized by means of the complementary conditions

$$x_{ij}^* > 0 \quad \Rightarrow \quad y_i + y_j - \sum_{\{i,j\} \in W} y_W = c_{ij}$$

$$y_W > 0 \quad \Rightarrow \quad x^*(\gamma(W)) = q_W.$$

Again an alternative to the above blossom–characterization of $P^{=}(G, b)$ is the following so–called *cut–characterization* where each blossom inequality for $W \in \mathcal{S}(b)$ is replaced by an equivalent *cut inequality*:

$$x(\delta(W)) \geq 1.$$

Thus the complete system of linear (in)equalities is:

$$x(\delta(i)) = b_i \quad \text{for } i \in V$$

$$x(\delta(W)) \geq 1 \quad \text{for } W \in \mathcal{S}(b)$$

$$x \geq 0,$$

and we get the following alternative dual (b–matching) problem

$$max \sum_{i \in V} b_i \cdot y_i + \sum_{W \in \mathcal{S}(b)} y_W \quad \text{subject to} \qquad \text{(DbMPII)}$$

$$y_i + y_j + \sum_{\{i,j\} \in \delta(W)} y_W \leq c_{ij} \quad \text{for } \{i,j\} \in E$$

$$y_W \geq 0 \quad \text{for } W \in \mathcal{S}(b).$$

And the associated complementary slackness conditions characterizing an optimal perfect b–matching x^* are

$$x_{ij}^* > 0 \quad \Rightarrow \quad y_i + y_j + \sum_{\{i,j\} \in \delta(W)} y_W = c_{ij}$$

$$y_W > 0 \quad \Rightarrow \quad x^*(\delta(W)) = 1.$$

Note that by $\bar{c}_{ij}$ we again denote the reduced cost of an edge $\{i,j\} \in E$ with respect to a dual solution of either characterization.

So far, only the blossom–characterization has been used algorithmically in primal–dual approaches (Pulleyblank [1973], Marsh [1979]). Yet it is evident how these "blossom–algorithms" can be reformulated to be based on the cut–characterization.

While the validity of the blossom–characterization was first proved algorithmically via the development of the primal–dual method (cf. Edmonds and Johnson [1970], Pulleyblank [1973]), the validity of the cut–characterization was recently shown in a constructive manner by Green-Krotki [1980] who first transforms the given (perfect) b–matching problem into its equivalent 1–matching problem by the reduction of section 5.3. and then derives the cut–characterization for $P^=(G,b)$ from the (known) cut characterization for the perfect 1–matching polytope.

For the 1–matching problem every matching M induces a <u>vertex</u> of the matching polytope P_M . However for the general b- matching problem this is not the case. Pulleyblank [1973] has characterized those b-matchings of G which are vertices of $P^\le(G,b)$. He proved his result by means of the primal-dual blossom algorithm.

Since for every vertex $x^0 \in P^\le(G,b)$ there exists a cost–vector (edge weights) $c_0 \in \mathbb{R}^E$ such that the bMP

$$min \ \{c_0'x \mid x \in P(G,b)\}$$

has x^0 as a <u>unique</u> solution and since this solution is found by the blossom algorithm, the vertices of $P^\le(G,b)$ are precisely the b–matchings produced — or better producable — by the blossom algorithm.

With respect to a b–matching $x^0 \in P^\le(G,b)$ we denote $G^+(x^0)$ as the spanning subgraph of G the edge set $E^+(x^0)$ of which is defined by

$$E^+(x^0) := \{e \in E \mid x_e^0 > 0\}.$$

Theorem 14.8. *(Pulleyblank [1973])*
A b–matching x^0 is a vertex of $P^\le(G, b)$ iff each component $H = (V(H), E(H))$ of $G^+(x^0)$ satisfies the following conditions:

(i)   *H contains no even polygon;*

(ii) *H contains at most one x^0–deficient node;*

(iii) *if H contains more than one polygon then in any path joining two odd polygons there is an edge $e \in E(H)$ — called isthmus — for which $x_e^0 = 1$;*

(iv) *if H contains a x^0–deficient node v and an odd polygon then either v has deficiency 1 or else for any odd polygon P contained in H there is an edge $e \in E(H)$ — called isthmus — for which $x_e = 1$ in any path from v to P .*

Given a b–matching $x \in P^{\leq}(G,b)$ we can transform x into a vertex x^0 by some simple combinatorial operations (modifications). Before we introduce these operations we reformulate the above characterization.

Corollary 14.9.

A b–matching x^0 is a vertex of $P^{\leq}(G,b)$ iff exists a shrinking family S in G such that each component $H = (V(H), E(H))$ of $(G \times S)^+(x^0)$ satisfies the following properties

(i) *H contains no even polygon and at most one odd polygon,*

(ii) *H contains at most one x_0–deficient node,*

(iii) *if H contains a polygon then $x(\delta(v)) = b_v$ for all $v \in V(H)$.*

In the following we will show how given any b–matching $x \in P^{\leq}(G,b)$ a b–matching x^0 and a shrinking family S can be constructed such that the above properties hold and

$$d(x, G, b) \geq d(x^0, G, b).$$

For that purpose assume $G \times S$ a surface graph with S a (possibly empty) shrinking family in G and x a b–matching in $G \times S$.

$\underline{b\text{–matching modification I}}$ ("even polygon deletion")

Let $P = (V(P), E(P))$ be an even polygon in $(G \times S)^+(x)$ then define

$$\epsilon := min\{x_e \mid e \in E(P)\}.$$

Let e_0 such that $x_{e_0} = \epsilon$ and M_0 the perfect 1–matching in P containing e_0..
Then set

$$x'_e := \begin{cases} x_e - \epsilon, & \text{for } e \in M_0 \\ x_e + \epsilon, & \text{for } e \in E(P) \setminus M_0 \\ x_e, & \text{otherwise.} \end{cases}$$

After this transformation $(G \times S)^+(x')$ does not contain the even polygon P, no new even polygon is constructed and $d(G, b, x') = d(G, b, x)$.

b–matching modification II

Let $H = (V(H), E(H))$ be a component of $(G \times S)^+(x)$ containing two x–deficient nodes u and v . W.l.o.g. assume $\epsilon_1 := d_u(x) \leq d_v(x)$. Let $P(u, v)$ be a simple path in H from u to v . Define

$$\epsilon_2 := min \; \{x_e \mid e \text{ an even edge of } P(u, v)\} \text{ and}$$

$$\epsilon := min \; \{\epsilon_1, \epsilon_2\}$$

and set

$$x'_e := \begin{cases} x_e + \epsilon, & \text{for } e \text{ odd edge of } P(u, v) \\ x_e - \epsilon, & \text{for } e \text{ even edge of } P(u, v) \\ x_e, & \text{otherwise.} \end{cases}$$

After this transformation either

(i) $d_u(x') = 0$,

(ii) an odd polygon P in H is destroyed or

(iii) u and v are no longer in the same component.

By this modification the deficiency of x' may become less than the deficiency of x ; this is the case if $P(u, v)$ has an odd number of edges.

b–matching modification III

Let $H = (V(H), E(H))$ be a component of $(G \times S)^+(x)$ with one x–deficient node u and an odd polygon $P = (V(P), E(P))$.

Let $P(u, w)$ be the simple path in H from u to the node $w \in V(P)$ which is closest to u and let Q be the (nonsimple) path in H obtained when first traversing $P(u, w)$ from u to w and then the odd polygon P. Define

$$\epsilon_1 := min\{x_e \mid e \text{ is an even edge of } Q, e \notin E(P)\}$$

$$\epsilon_2 := min\{x_e \mid e \text{ is an even edge of } Q, e \in E(P)\}$$

$$\epsilon := min\{\lfloor \epsilon_1/2 \rfloor, \epsilon_2, \lfloor d_u(x)/2 \rfloor\}.$$

(Note that $u = w$ is possible in which case $Q \subseteq E(P)$ and hence $\epsilon_1 = \infty$.)

Then perform the following transformation (with $\epsilon = 0$ possibly).

$$x'_e := \begin{cases} x_e + 2\epsilon, & e \text{ odd edge of } Q, e \notin E(P) \\ x_e + \epsilon, & e \text{ odd edge of } Q, e \in E(P) \\ x_e - \epsilon, & e \text{ even edge of } Q, e \in E(P) \\ x_e - 2\epsilon, & e \text{ even edge of } Q, e \notin E(P) \\ x_e, & \text{otherwise.} \end{cases}$$

After this transformation at least one of the following cases must occur:

(i) $d_u(x') = 0$,

(ii) P is no longer contained in $(G \times S)^+(x')$,

(iii) u and P are no longer in the same component of $(G \times S)^+(x')$

(iv)   there exists an edge $e \in P(u, w)$ with $x'_e = 1$ or

(v) $d_u(x') = 1$.

In the first three cases we have reduced the number of components H in $(G \times S)^+(x')$ with the property that H contains a x'-deficient node and an odd polygon. In the latter cases we have discovered a blossom $B = (V(B), E(B))$ with respect to x', which we will shrink. If (iv) holds, B is the part of H containing P after removing e, and if (v) holds, the whole component forms the blossom B.

Define $S := (V \cap V(B)) \cup (\cup\{R \mid R \in V(B) \cap \mathcal{S}\})$ and set $\mathcal{S} := \mathcal{S} \cup \{S\}$. Then the component of $(G \times S)^+(x')$ containing pseudonode S does not contain a polygon.

Using these b–matching modifications successively, after a finite number of modifications, a pair (x, S) is constructed such that $(G \times S)^+(x)$ fulfills the properties of Corollary 14.9. . The finiteness can be seen as follows: Any application of one of the three modifications either reduces the number of edges e with $x_e \neq 0$ by at least one or reduces the number of x–deficient nodes by at least one or shrinks a subgraph of at least three nodes. Thus after at most

$$K := |\{e \in E \mid x_e > 0\}| + |\{v \in V \mid d_v(x) < b_v\}|$$

applications of either modification the procedure stops.

The amount of work in a single modification step is bounded by $O(|E|)$, thus the total amount is bounded by $O(|E|^2)$.

14.3. Alternating Trees and Augmentations

For the 1–matching problem an augmenting path with respect to a matching M is a (simple) path connecting two M–unsaturated nodes with the edges alternately in M and not. Then exchanging the role of matching and non-matching edges along P a matching M' is obtained with $|M'| = |M| + 1$. The basic idea of "improving" matchings by alternating paths can be extended to the b–matching problem, too. This is easily seen by transforming bMP into the equivalent 1MP since we know that a b–matching in G is of maximum cardinality iff the associated 1–matching is a maximum cardinality 1–matching.

Yet an augmenting path with respect to the associated 1–matching may lead to a non–simple (augmenting) path with respect to the b–matching in G .

The following figure demonstrates the complicating possibility of non–simple augmenting paths with respect to a b–matching. (Here the nodes carry their b–value and the edges carry the current x–value.)

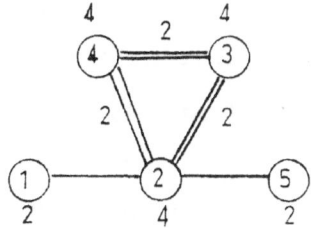 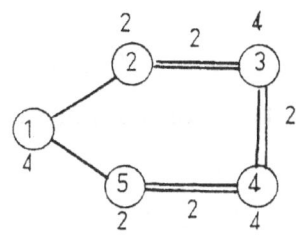

Figure 14.2.: Two non–simple augmenting paths

In the first example the sequence

$$P = (\{1,2\}, \{2,3\}, \{3,4\}, \{4,2\}, \{2,5\})$$

describes an augmenting path while for the second example

$$P = (\{1,2\}, \{2,3\}, \{3,4\}, \{4,5\}, \{5,1\})$$

induces an augmenting path.

The following figure shows the "improved" b–matching.

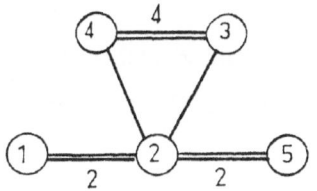 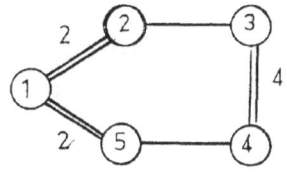

Figure 14.3. Improved b–matchings

In the following we develop the basic routines for constructing maximum cardinality b–matchings. For that purpose we assume x a b–matching which is a vertex of $P^{\leq}(G, b)$ and a surface graph $G \times S$ with S a shrinking family in G such that the properties of Corollary 14.9. hold for $(G \times S)^{+}(x)$. This assumption is not restrictive since given any b–matching x^0 , a vertex x of $P^{\leq}(G, b)$ with $d(G, b, x^0) \geq d(G, b, x)$ can be constructed in polynomial time by means of the three b–matching modification procedures. Moreover the trivial start solution $x \equiv 0$ fulfills this assumption.

The routines presented here are also essential ingredients for the primal–dual blossom–algorithm where successively maximum cardinality b–matchings have to be constructed and in the shortest augmenting path algorithm, too.

To ease notation we will denote the surface graph on which our procedure is working by $G = (V, E)$ again. Hence some of the nodes $i \in V$ may be pseudonodes and some of the edges in E may connect different "real" nodes contained in different pseudonodes. In any case we assume that the properties of Corollary 14.9. hold for G with respect to the b–matching x in G .

A tree $T = (V(T), E(T))$ in G rooted at a x–deficient node r is called x–alternating iff

(i) $x(\delta(i)) = b_i$ for $i \in V(T) \setminus \{r\}$

(ii) $x(\delta(V(T))) = 0$

(iii) $x_e > 0$ for every even edge of T.

A nonempty collection F of node-disjoint x–alternating trees is called a x–alternating forest if every x–deficient node is the root of one of the trees in F .

Again x–alternating tress lead to a bicoloring of their nodes into *outer nodes* resp. *inner nodes* which have the same color as the root resp. not the same color as the root and these sets are denoted by $O(T)$ and $I(T)$, respectively.

As in the 1-matching case given a x-alternating forest $F = (V(F), E(F))$ we would "*branch from an outer node $i_1 \in V(F)$ into an edge $\{i_1, i_2\}$ not yet contained in $E(F)$*". Then we can distinguish the following cases:

Case 1 ("two–tree–augmentation")

As in the 1-matching case a "x-augmenting path" $P = (V(P), E(P))$ exists if i_1 and i_2 are outer nodes from different trees, T_1 and T_2 say.

Let r_1 resp. r_2 be the roots of these trees and let P_1 resp. P_2 be the paths from r_1 to i_1 in T_1 and from r_2 to i_2 in T_2 , respectively. Then calculate

$$\epsilon_1 := min\{x_e \mid e \text{ is an even edge of } P_1 \text{ resp. } P_2\}$$
$$\epsilon_2 := min\{d_{r_1}(x), d_{r_2}(x)\}$$
$$\epsilon := min\{\epsilon_1, \epsilon_2\}$$

and define a b-matching x' in the following way

$$x'_e := \begin{cases} x_e + \epsilon, & \text{if } e \text{ is an odd edge of } P_1 \text{ resp. } P_2 \text{ or } e = \{i_1, i_2\} \\ x_e - \epsilon, & \text{if } e \text{ is an even edge of } P_1 \text{ resp. } P_2 \\ x_e, & \text{otherwise.} \end{cases}$$

Then $d(G, b, x') = d(G, b, x) - 2\epsilon$. The above operation is called the *two–tree–augmentation* via $P = (P_{i_1}, \{i_1, i_2\}, P_{i_2})$ and the value ϵ is called the *capacity* of P and we write shortly

$$x' := x \bigoplus \epsilon \cdot x(P).$$

For $i = 1, 2$, if $x'(\delta(r_i)) = b_i$, remove T_i from F else let f_i be the first even edge on the path P_i for which $x'_{f_i} = 0$. If such an edge exists, remove f_i and the portion of T_i not containing r_i after removing f_i to obtain T'_i .

By the choice of ϵ at least one of these cases must occur and hence at most one of i_1 and i_2 are in the new forest F' . If neither are in F' set $F := F'$.

If one, i_1 say, belongs to F , then edge $\{i_1, i_2\}$ connects an outer node and a node not contained in the x-alternating forest. Let H_2 be the component of

275

$G^+(x')$ containing i_2 . H_2 is a tree, T_2' say. Now "grow" T_1' , the x'–alternating tree containing i_1 , by attaching tree T_2' by means of edge $\{i_1, i_2\}$ and identify the new outer resp. inner nodes in T_1' . Set $x := x'$ and $F := F'$. This completes the one–tree–augmentation.

A different situation occurs if i_1, i_2 are outer nodes of the same x–alternating tree T rooted at r . In that case let P_1 resp. P_2 be the paths from r to i_1 in T resp. from r to i_2 in T , and let $\bar{P}$ be the common portion of P_1 and P_2 and $\bar{P}_1$ and $\bar{P}_2$ the respective remaining portions of P_1 and P_2 .

Then $\bar{P}_1, \bar{P}_2$ and the edge $\{i_1, i_2\}$ form an odd cycle. Calculate

$$\epsilon_1 := 1/2 \, min\{x_e \mid e \text{ is an even edge of } \overline{P}\}$$

$$\epsilon_2 := min\{x_e \mid e \text{ is an even edge of } \overline{P}_1 \text{ resp. } \overline{P}_2\}$$

$$\epsilon_3 := 1/2 d_r(x)$$

$$\epsilon(i_1, i_2) := min\{\lfloor \epsilon_1 \rfloor, \epsilon_2, \lfloor \epsilon_3 \rfloor\}.$$

Let us call the above calculation the "ϵ–CHECK" for edge $\{i_1, i_2\}$.

Case 2 ("one–tree–augmentation")

If i_1, i_2 are outer nodes of the same tree T rooted at r and $\epsilon(i_1, i_2) > 0$, then the non–simple path P obtained by traversing $\bar{P}, \bar{P}_1, \{i_1, i_2\}, \bar{P}_2, \bar{P}$ starting and ending at r is an "x–augmenting path", too. The new b–matching x' is defined as follows with $\epsilon := \epsilon(i_1, i_2)$

$$x_e' := \begin{cases} x_e + 2\epsilon, & \text{if } e \text{ is an odd edge of } \bar{P} \text{ or } e = \{i_1, i_2\} \\ x_e - 2\epsilon, & \text{if } e \text{ is an even edge of } \bar{P} \\ x_e + \epsilon, & \text{if } e \text{ is an odd edge of } \bar{P}_1 \text{ resp. } \bar{P}_2 \\ x_e - \epsilon, & \text{if } e \text{ is an even edge of } \bar{P}_1 \text{ resp. } \bar{P}_2 \\ x_e, & \text{otherwise.} \end{cases}$$

Then $d(G, b, x') = d(G, b, x) - 2\epsilon$ and we call the above operation the one–tree augmentation via $P = (\bar{P}, \bar{P}_1, \{i_1, i_2\}, \bar{P}_2, \bar{P})$. Again ϵ is called the capacity of

P and we write shortly

$$x' = x \bigoplus \epsilon \cdot x(P).$$

Then set $x := x'$ and if r is not x–deficient any more let F' be the forest obtained from F by removing T and set $F := F'$.

If $x(\delta(r)) < b_r$, then the component H_r of $G^+(x)$ containing r may contain a deficient node and an odd polygon. Thus the properties of Corollary 14.9. are no longer fulfilled. The following modifications reconstruct the x–alternating forest such that these properties hold again.

(i) If there are edges $f \in E(\bar{P})$ such that $x_f = 0$, let f^* be the first such edge. Remove f^* and the portion of T not containing r after removing f^* to obtain the new forest F^* . Set $F := F^*$.

(ii) If $x_f > 0$ for all $f \in E(\bar{P})$ but $x_f = 0$ for some edges $g \in E(\bar{P}_1) \cup E(\bar{P}_2)$, then remove these edges to obtain a new forest F^* . Set $F := F^*$.

(iii) If $x_h > 0$ for all $h \in E(P) = E(\bar{P}) \cup E(\bar{P}_1) \cup E(\bar{P}_2) \cup \{\{i_1, i_2\}\}$) then $T = (V(T), E(T))$ is a blossom with respect to x . Shrink T to a pseudonode. In the new surface graph $G' = G \times \{T\}$ the pseudonode T is an outer node with $b_T = 1$. Let x' be the b–matching in G' induced by x . Set $x := x'$ and set $G := G'$.

(Note, if G is a surface graph already, i.e. $G = G \times S$, set

$$S := (V(T) \cap V) \cup \bigcup \{R \mid R \in V(T), R \in S\} .$$

Then set $S' := S \cup \{S\}$, set $b_S := 1$ and let x' be the matching in $G \times S'$ induced by x . Set $S := S'$ and set $x := x'$.)

If i_1, i_2 are outer nodes of the same tree T rooted at r but the ϵ–CHECK, yields $\epsilon(i_1, i_2) = 0$, then no one–tree augmentation is possible and we have detected a blossom with respect to x which we will shrink now.

(This step is very similar to the operations necessary after a one–tree augmentation and we use the same notation as in Case 2.)

Let w be the first even node of $\bar{P}$ such that the portion $\tilde{P}$ of $\bar{P}$ connecting w and the odd polygon in T does not contain an even edge e with $x_e = 1$. Then the blossom $B = (V(B), E(B))$ consists of the subgraph of T induced by $\tilde{P}$ and the odd polygon except for the even edge of T incident with w, if it exists. As in Case 2, this blossom is now shrunk to a pseudonode.

Case 4 ("tree–growing")

If $i_2 \notin V(F)$, i.e. i_2 is not yet contained in any x–alternating tree, let H_2 be the component of $G^+(x)$ containing i_2.

If H_2 does not contain a polygon, then H_2 is a tree, T_2 say. Otherwise H_2 contains an odd polygon, $P = (V(P), E(P))$ say. In that case let $k \in V(P)$ such that the path from i_2 to k in H_2 is odd and let $f \in E(P)$ be an edge joining k to a node l, which is not closer to i_2 than k. Then let T_2 be the tree obtained from H_2 by deleting edge f.

In either case grow T, the tree containing i_1, by attaching tree T_2 by means of edge $\{i_1, i_2\}$ and identify the new outer resp. inner nodes in T. This operation is called "growing T via edge $\{i_1, i_2\}$".

If the tree T_2 was obtained by deleting an edge f from an odd polygon, then this edge now joins two outer nodes of T. Thus we will now perform an ϵ–CHECK for edge f and either perform a "one–tree augmentation" or a "blossom–shrinking". In both cases a x–alternating forest will be constructed finally fulfilling the properties of Corollary 14.3. .

From the above discussion of the several possible cases, it becomes clear that there are a number of interdependencies among the different routines to guarantee that the properties of Corollary 14.9. are maintained throughout the procedure, or more precisely, are retained after a possible sequence of operations. So the occurence of <u>case 1</u> and the following one–tree augmentation may require the performance of a subsequent special grow–step etc. .

In the course of the cardinality b–matching algorithm (and the primal–dual b–matching algorithms) x–alternating forests are "grown" or better "manipulated" via the above operations. As already mentioned, we thereby work on a surface graph $G \times S$. Now analogously to the 1–matching case we call a x–alternating forest F in $G \times S$, with S a possibly empty shrinking family in G , a Hungarian forest iff

(i)   all outer nodes are joined to inner nodes only

(ii)   all pseudonodes are non–inner nodes.

Then the following lemma generalizes Lemma 9.6. .

Lemma 14.10.

Let F be a Hungarian forest in $G \times S$ with respect to the b–matching x_S in $G \times S$. Then x_S induces a maximum cardinality b–matching x in G with $x(\gamma(R)) = q_R$ for $R \in S$.

Proof. (similar to the 1-matching case, cf. Marsh [1979] for instance.)

□

Let us now discuss the complexity of the subroutines and the entire cardinality b–matching algorithm.

A bound of $O(|E|)$ applies to the work in a "tree–growing", "blossom–shrinking" or "augmentation"–step. An alternating forest can be grown at most $|V|$ times without either an augmentation or without producing a Hungarian forest. Shrinking can be performed at most $(|V|-1)/2$ times. Thus after at most $O(|V|\cdot|E|)$ steps an augmentation occurs, decreasing $d(G\times S,b,x)$ by at least two, or a Hungarian forest is established, thus showing that the current b–matching is of maximum cardinality.

Hence starting with an initial b-matching x_0 in G, after at most

$$O(d(G,b,x_0)\cdot|V|\cdot|E|)$$

operations a maximum cardinality b–matching is obtained.

Note that when "starting" the cardinality b–matching procedure with $x\equiv 0$ and $S=\emptyset$, then throughout the procedure all pseudonodes which are created are non–inner nodes of the alternating forest.

Yet as in the 1–matching case, also in algorithms for solving (weighted) b–matching problems we may be forced to "*expand*" inner pseudo-nodes of an x_S–alternating forest F in $G\times S$.

For reasons of completeness we present this expansion–routine already in this context:

Pseudonode expansion

Assume $R\in S$ an inner pseudonode in $G\times S$ and let $B[R]$ be the b–critical pseudotree spanning $G[R]\times S[R]$. Let $e\in E$ be the unique edge of the alternating forest incident with R such that $x_e=1$ and let $i\in R$ be the node of R which is incident with e.

Construct a near–perfect matching x' of $B[R]$ which is deficient at i. Set $S:=S\setminus\{R\}$ and extend x_S using x'.

(i) If e is an odd edge of the tree T , then it is the only edge of T incident with R . Let T' be the tree obtained by replacing R by the component of $(G \times S)^+$ containing i .

(ii) If e is an even edge of T , then let f be the unique odd edge of T incident to R and let j be the node of R incident with f . Let $P(i,j)$ be the simple path in $B[R]$ from i to j having even length. Then let T' be the tree obtained from T by replacing R with $P(i,j)$ and any component of $(G \times S)^+$ which contains a node of $P(i,j)$.

In either case the new "tree" T may contain the odd polygon $P(B[R])$. In that case let w be a node of $P(B[R])$ at odd distance from j with minimal distance. Let t be a node of $P(B[R])$ adjacent to w and not belonging to the odd–length path in $B[R]$ joining j to w . Let g be the edge joining w and t . Remove g from T' and set $T := T'$. Identify the new outer resp. inner nodes in T .

14.4. Extreme b–Matchings

For the 1–matching problem we have characterized extreme matchings via the lack of negative alternating cycles. A similar definition is also possible for b–matchings by means of transforming the b–matching problem on $G = (V, E)$ into the equivalent 1–matching problem on $\tilde{G} = (\tilde{V}, \tilde{E})$.

Thus according to this transformation a b–matching x in G would be called extreme if the associated 1–matching M_x in $\tilde{G}$ is extreme, i.e. does not allow negative M_x–alternating cycles. Yet, we call such b–matchings *pseudo–extreme* only, and for a pseudo-extreme b–matching to be extreme we will require an additional property.

It is obvious that any pseudo–extreme b–matching x is an optimal (perfect)

b–matching for the bMP with altered node constraints

$$b'_i = x(\delta(i)) \text{ for } i \in V$$

and vice versa.

Let $S(b') = \{R \subseteq V \mid R \quad b' - \text{critical}\}$ then exists vector $y \in \mathbb{R}^{V \cup S(b')}$ which is dual feasible with respect to the blossom–characterization resp. cut–characterization of $P^=(G, b')$ and has the following properties

$$(i) \quad x_{ij} > 0 \Rightarrow \bar{c}_{ij} = 0 \text{ for } \{i, j\} \in E$$

$$(ii) \quad y_W > 0 \Rightarrow x(\gamma(W)) = (b'(W) - 1)/2 \text{ and }$$

$$x(\delta(W)) = 1 \text{ for } W \in S(b').$$

And vice versa, given such a vector y shows that x is an optimal b'–matching, hence x is a pseudo–extreme b–matching.

Note that $S(b)$ and $S(b')$ might be completely "unrelated" i.e.

$$S(b) \not\subseteq S(b') \text{ and } S(b') \not\subseteq S(b).$$

Thus a "dual solution" $y' \in \mathbb{R}^{V \cup S(b')}$ cannot be extended to a dual solution $y \in \mathbb{R}^{V \cup S(b)}$ in general.

Given a b–matching problem on a graph $G = (V, E)$ and a vector $y \in \mathbb{R}^{V \cup S(b)}$ which is dual feasible with respect to either characterization we define

$$E(y) := \{\{i, j\} \in E \mid \bar{c}_{ij} = 0\} \text{ and } G(y) = (V, E(y))$$
$$S(y) := \{R \in S(b) \mid y_R = 0\} \text{ and }$$
$$\overline{S(y)} := S(b) \setminus S(y).$$

Moreover, for any b–matching x we define $(G \times \overline{S(y)})^+$ to be the (spanning) subgraph of $G \times \overline{S(y)} = (\bar{V}, \bar{E})$ having edge set $E^+ := \{e \in \bar{E} \mid x_e > 0\}$.

Now a pair (x, y) with x a b–matching in G and $y \in \mathbb{R}^{V \cup S(b)}$ is called a *strongly compatible pair* iff

(i) $x_e > 0 \Rightarrow e \in E(y)$ for $e \in E$

(ii) $R \in \overline{S(y)} \Rightarrow x(\delta(R)) = 1$ and $x(\gamma(R)) = q_R$

(iii) $\overline{S(y)}$ is a shrinking family in $G(y)$.

Then the following lemma holds.

Lemma 14.11.

Let (x, y) be a strongly compatible pair, then x is a pseudo–extreme b–matching.

Proof.   From property (ii) follows that $R \in \overline{S(y)}$ implies $x(\delta(i)) = b_i$ for all
$i \in R$. Now define $b' \in \mathbb{N}_0^V$ via $b_i' = x(\delta(i))$ for $i \in V$ then $R \in \overline{S(y)}$ implies
$R \in S(b')$ thus $\overline{S(y)} \subseteq S(b')$. Now define

$$y_i' = y_i \quad \text{for} \quad i \in V$$

$$y_R' = \begin{cases} y_R, & \text{for } R \in \overline{S(y)} \\ 0, & \text{for } R \in S(b') \setminus \overline{S(y)}. \end{cases}$$

Then y' is a feasible solution for the dual of $P^=(G, b')$ which together with
$x \in P^=(G, b')$ fulfills the complementary slackness conditions. Hence x is an
optimal perfect b'–matching and therefore a pseudo–extreme b–matching.

$\square$

Any pseudo–extreme b–matching x is a feasible point in $P^{\leq}(G, b)$. Yet
x need not be an extreme point (vertex) of $P^{\leq}(G, b)$. Now define a pseudo–
extreme b-matching x to be *extreme* iff it is a vertex of $P^{\leq}(G, b)$. Thus we can
characterize extreme b–matchings x by the following properties

(i)   Exists $y \in \mathbb{R}^{V \cup S(b)}$ such that (x, y) strongly compatible.

(ii)   Exists a shrinkable family $S \supset \overline{S(y)}$ such that each component $H =$
$(V(H), E(H))$ of $(G \times S)^+(x)$ fulfills the following properties

– H contains at most one x–deficient node,

– H contains no even polygon and at most one odd polygon,

– If H contains an odd polygon, then $x(\delta(i)) = b_i$ for $i \in V(H)$.

Given an extreme b–matching x in G we call (x, y, S) fulfilling the above properties a *strongly compatible triple* and we define x_S as the b–matching in $G \times S$ induced by x .

Note that using the matching modification routines of section 14.2. any pseudo–extreme b–matching $\tilde{x}$ can be transformed into an extreme b–matching x with

$$d(G, b, x) \leq d(G, b, \tilde{x}).$$

Thereby the shrinking family S is constructed, too.

Let P be a x–augmenting path and ϵ the capacity of P . With $\tilde{x} := x \oplus \epsilon \cdot x(P)$ we define

$$c(P) := 1/\epsilon \cdot (c'\tilde{x} - c'x)$$

the *length* of P .

Note that we can also define the length of P in the following way

$$c(P) = \sum\{c_e \mid \tilde{x}_e = x_e + \epsilon\} - \sum\{c_e \mid \tilde{x}_e = x_e - \epsilon\}.$$

Now the following lemma holds obviously.

Lemma 14.12.

Let x be a *pseudo–extreme* b–*matching and* P *a shortest* x–*augmenting path with capacity* ϵ . *Then for all* $1 \leq \delta \leq \epsilon$

$$x' := x \oplus \delta \cdot x(P)$$

is a pseudo–extreme b–*matching.*

Proof. Follows from the fact that P induces a shortest augmenting path Q with respect to the associated 1-matching M_x in $\tilde{G}$. Hence$M_{x'} = M_x \oplus Q$ is extreme in $\tilde{G}$ and thus $x' = x \oplus 1 \cdot x(P)$ is pseudo-extreme in G.

□

Moreover the following lemma holds:

Lemma 14.13.

Let (x, y, S) be a strongly compatible triple and P a shortest x-augmenting path with capacity ϵ which is induced by an augmenting path in $G \times S$. Then

$$x' := x \oplus \epsilon \cdot x(P)$$

is an extreme b–matching again.

Proof. Due to the above lemma, we only have to show that x' is a vertex of $P^{\leq}(G, b)$ again. For that purpose we only have to show the existence of a dual solution $y' \in \mathbb{R}^{V \cup S(b)}$ and a shrinking family $S' \supset \overline{S(y)}$ such that (x, y, S') is a strongly compatible triple. Now the existence of y' follows from the fact that x' is pseudo–extreme and the existence of a suitable shrinking family S can be followed from the results of section 14.3. and the operations in connection with an augmentation.

□

Thus as in the 1–matching case, optimal perfect b–matchings can be constructed by sequentially augmenting along shortest augmenting paths. Starting from $x_0 \equiv 0$, $y \equiv 0$ and $S = \emptyset$ at most $d(G, b, x_0)/2 = (\sum_{i \in V} b_i)/2$ such augmentations are necessary.

It remains to show how shortest augmenting paths relative to an extreme matching x can be constructed.

Given a strongly compatible triple (x, y, S) and the b–matching x_S in $G \times S$, the following analoga to the 1–matching case hold.

Lemma 14.14.

Let P be a x-augmenting path which is induced by a x_S- augmenting path in $G \times \overline{S(y)}$ joining two x-deficient nodes s and t . Then

$$c(P) = \bar{c}(P) + y_s + y_t.$$

Lemma 14.15.

Let P be a x-augmenting path and let R be a maximal member of $\overline{S(y)}$. For P_R the portion of P contained in R we get

$$\bar{c}(P_R) = 0.$$

Lemma 14.16.

Let P be a x-augmenting path joining two x-deficient nodes s and t which is not induced by a x_S-augmenting path in $G \times S$ but which is induced by an augmenting path in $G \times (S \setminus \{R\})$ with R a maximal member of S . Then

$$c(P) = \begin{cases} \bar{c}(P) + y_R + y_s + y_t, & \text{for the blossom charcterization} \\ \bar{c}(P) + 2y_R + y_s + y_t, & \text{for the cut-characterization.} \end{cases}$$

The proofs of these lemmata are similar to the proofs in the 1–matching case.

As in the 1–matching case these lemmata motivate to determine a shortest x_S–augmenting path in $(G \times \overline{S(y)})$ with respect to the reduced cost coefficients and to expand inner pseudonodes if "there is evidence" that the shortest x–augmenting path in G might not be induced by the shortest x_S– augmenting path in $G \times S$.

Chapter 15. b–Matching Algorithms

So far we have introduced the combinatorial structures which are relevant when working with b–matchings in graphs and we have extended the concept of extreme matchings to the b–matching case. Again this concept builts the basis for the shortest augmenting path approach.

In this chapter we will first outline how the SAP–approach can be implemented in the b–matching case. Then we will show how this approach combined with the scaling–technique of Edmonds and Karp leads to a polynomial b-matching algorithm.

15.1. The Shortest Augmenting Path Approach

The SAP–algorithm for solving (perfect) b–matching problems is similar to the SAP–labeling method for solving 1MP. Again we can consider two variants depending on the characterization of $P^{=}(G, b)$ which is used.

In the following we will shortly present the method based on the cut–characterization. Thus we assume (x, y, S) a strongly compatible triple with x_S the b–matching in $G \times S$ induced by x . As in the 1MP-method we assign labels d_i^+ and d_i^- to the (pseudo–)nodes in $G \times S$ building up a shortest x_S–alternating path tree with respect to the reduced cost–coefficients. A central operation is to "scan" outer (pseudo–)nodes. This subroutine is the same as in the 1MP–algorithm, i.e. if the outer node is a pseudonode $R \in S$ we have to examine all edges in $\delta(R)$. We will not formulate the SCAN–subroutine explicitly here, since we only want to outline the SAP–algorithm for solving bMP, thereby avoiding the rather technical notation (definition of $b(i)$, $B(i)$ etc.). Let us define by $\bar{V}$ the set of (pseudo–)nodes in $G \times S$, and by $\bar{E}$ the set of edges in $G \times S$.

The procedure is initialized by the following:

INITIALIZATION

Set $d_i^+ := d_i^- := \infty$ for all $i \in \bar{V}$.

For each tree–component H of $(G \times S)^+(x)$ containing a x–deficient

node r make r the outer root of this tree and identify the outer and inner

nodes in $\bar{V}(H)$ and set

$$d_i^+ := 0 \text{ for } i \in \bar{V}(H) \text{ outer}$$
$$d_i^- := 0 \text{ for } i \in \bar{V}(H) \text{ inner} .$$

SCAN all outer nodes in $V(H)$. This gives a x–alternating forest F .

Then the whole procedure is again controlled by the following minimization– and branching–routine.

CONTROL

Calculate

$$\delta_1 : = min\{d_i^- \mid i \notin \bar{V}(F)\}$$
$$\delta_2 : = min\{1/2(d_i^- + d_i^+) \mid i \in \bar{V}(F), \text{ outer }\}$$
$$\delta_3 : = min\{d_R^- + y_R \mid R \in \bar{V}(F) \text{ inner pseudonode }\}.$$

If all the above sets are void, then F is a Hungarian forest in $G \times S$,

hence no perfect b–matching exists, STOP.

Set $\delta := min\{\delta_1, \delta_2,\}$ and let $i_0 \in \bar{V}$ be the (pseudo–)node defining δ .

If $\delta_3 < \delta$ let R_0 be the pseudonode defining δ_3 .

288

If $\delta_3 < \delta$ $\rightarrow$ EXPAND(R_0)

If $\delta = \delta_1$ $\rightarrow$ GROW(i_0)

If $\delta = \delta_2$ $\rightarrow$ CHECK(i_0)

In the case that $\delta = \delta_2$, as in the 1MP–case, either an augmentation is possible or a blossom has been detected which will be shrunk. Yet for bMP there are two different types of augmentations possible. Now the following routine CHECK(i_0) will determine which of these altogether three possible cases applies. Thereby we will make use of the (combinatorial) routine ϵ–CHECK which has been introduced in section 14.3. and which determines whether a one–tree–augmentation is possible.

CHECK (i_0)

Let j_0 be the (pseudo-)node in $\bar{V}$ defining $d_{i_0}^-$.

If i_0 and j_0 are contained in different trees

$\rightarrow$ TWO–TREE AUGMENT (i_0, j_0),

if i_0 and j_0 are contained in the same tree perform ϵ – CHECK (i_0, j_0).

If $\epsilon(i_0, j_0) > 0$ $\rightarrow$ ONE–TREE AUGMENT (i_0, j_0)

If $\epsilon(i_0, j_0) = 0$ $\rightarrow$ SHRINK (i_0, j_0)

The "graphical" operations to be performed in the subroutines called by CONTROL resp. CHECK(i_0) have been introduced already in connection with the cardinality b–matching algorithm. Thus here we only state the additional operations involving the calculation of path–lengths.

<u>GROW (i_0)</u>

Let j_0 be the (pseudo-)node in $\bar{V}$ defining $d_{i_0}^-$ and let T be the alternating tree containing j_0.

Grow T via edge $\{j_0, i_0\}$.

For all new outer resp. inner (pseudo–)nodes j in T set

$$d_j^+ := \delta \text{ resp. } d_j^- := \delta$$

and SCAN all new outer nodes.

Return to CONTROL.

(If the tree T in GROW(i_0) was obtained by deleting an edge from the component containing i_0 , then this edge now joins two outer (pseudo–)nodes $,l$ in $\bar{V}(T)$ with

$$d_k^+ + d_k^- = 2\delta .$$

hus the next step is to perform ϵ-CHECK(k, l).)

ote that in the graphical operations of the following two routines the shrinking mily S has changed.

<u>SHRINK (i_0, j_0)</u>

Make the new pseudonode S an outer node of the alternating forest

and SCAN all formerly inner (pseudo–)nodes in S .

Set $y_S := 0$ and $d_S^+ := \delta$.

Return to CONTROL.

EXPAND (R_0)

Set $\delta := \delta_3$.

For all new outer resp. inner (pseudo–)nodes $i \in R_0$

set $d_i^+ := \delta$ resp. $d_i^- := \delta$.

SCAN all new outer (pseudo–)nodes $i \in R_0$.

For all (pseudo–)nodes $i \in R_0, i \notin \bar{V}(F)$ set

$$d_i^- := min\{d_j^+ + \bar{c}_{ji} \mid j \in \bar{V}(F) \text{ outer }\}.$$

Return to CONTROL.

The routines ONE–TREE AUGMENT(i_0, j_0) and TWO–TREE AUGMENT(i_0, j_0) are completely described in section 14.3. already.

A one–tree augmentation or two–tree augmentation in $G \times S$ is now followed by a dual–update–procedure which ends the current phase.

DUAL-UPDATE

Set
$$y_i := \begin{cases} y_i + (\delta - d_i^+), & \text{for } i \in \bar{V}(F), \text{ outer} \\ y_i + (d_i^- - \delta), & \text{for } i \in \bar{V}(F), \text{ inner} \\ y_i, & \text{if } i \in \bar{V} \setminus \bar{V}(F). \end{cases}$$

STOP.

After having performed this dual–update a new strongly compatible triple (x, y, S) is at hand. This also shows the correctness of the above procedure.

Starting with a strongly compatible triple (x_0, y_0, S_0) at most $O(d(G, b, x_0))$ shortest augmenting paths have to be constructed. Each shortest augmenting path computation, i.e. performance of the above procedure — consumes $O(|V||E|)$ steps. Thus the shortest augmenting path method for solving b-matching problems is of complexity $O(\sum_{i \in V} b_i \cdot |V| \cdot |E|)$ and this is <u>not</u> a bound which is polynomial in the input–size of the problem.

As for min-cost flow problems resp. Hitchcock transportation problems the (pure) shortest augmenting path concept has to be combined with a special anticipant "decison rule" to become a procedure the amount of which grows only polynomially with the size of the input. Yet in contrast to the HTP–case some difficulties have to be overcome when applying the scaling technique for bMP. This will be demonstrated in the next section.

Finally we want to mention that the SAP–approach introduced above is again "near–equivalent" to the primal–dual blossom algorithm. This can be seen from the fact that in the blossom–implementation a sequence of strongly compatible triples (x, y, S) is constructed, the b-matchings x of which are obtained by successively augmenting along augmenting paths. Since (pseudo–)extreme b-matchings are maintained throughout the procedure, the augmenting paths must be shortest augmenting paths.

As in the 1MP–case one also can reformulate the update–formulas of the blossom–implementation to obtain the more economical SAP–updating formulas.

15.2. The Scaling Approach

Marsh [1979] was the first who presented a good i.e. polynomial b–matching algorithm by combining Pulleyblank's blossom–algorithm for solving bMP with Edmonds and Karp's scaling technique, originally developed for solving min–cost–flow problems.

In this section we develop the scaling approach as a special implementation of the SAP–approach. We will explicitly use the fact that b–matching problems with even degree constraints, i.e $b \equiv 0 \mod 2$, can be reduced to a Hitchcock transportation problem which can be solved in polynomial time by the SAP/scaling– approach.

Initially we define

$$b'_i = 2\lfloor b_i/2 \rfloor \quad \text{for } i \in V.$$

Then the b–matching problem (resp.b'–matching problem)

$$
\begin{aligned}
min \ c'x \quad &\text{subject to} \qquad\qquad (P_{b'}) \\
Ax &= b' \\
x &\geq 0 \text{ integer}
\end{aligned}
$$

with A the node–edge incidence matrix of the non–bipartite graph G may have no solution, although the original b–matching problem has feasible solutions.

Thus we intend to construct a (not necessarily perfect) maximum cardinality b'–matching of minimal cost. Let K be the cardinality of a maximum cardinality b'–matching i.e.

$$K = max\{x(E) \mid Ax \leq b', x \text{ integer } \}$$

then this problem is denoted by $P_{b',K}$ and it can be formulated in the following way

$$\min \ c'x \quad \text{subject to} \qquad\qquad (P_{b',K})$$

$$Ax \leq b'$$

$$1'x = K$$

$$x \geq 0 \text{ integer.}$$

Problem $P_{b',K}$ can also be described as finding an extreme b'–matching x_0 in $P^{\leq}(G, b')$ of maximum cardinality. Then it is evident that given such an optimal extreme b'–matching x^0 together with the associated dual solution y^0 and shrinking family S^0 , the shortest augmenting path method for determining an optimal b–matching can be "started" from this strongly compatible triple (x^0, y^0, S^0) . Now the following lemma comprises the essential properties of this procedure.

Lemma 15.1.

(i) $P_{b',K}$ can be transformed into an equivalent HTP which can be solved in $O(\log \sum_{i \in V} b'_i \cdot |V| \cdot |E|)$ steps.

(ii) From the optimal HTP–solution a strongly compatible pair (x^0, y^0) with $\overline{S(y)} = \emptyset$ can be constructed.

(iii) $d(G, b, x^0) \leq |V|$.

Proof.   ad (i): With $P_{b',K}$ we associate the complete bipartite graph $\tilde{G} = (V_s, V_t, \tilde{E})$ where V_s and V_t are two copies of V . The cost–function (edge–weights) is defined as follows

$$\tilde{c}_{i_s j_t} := \tilde{c}_{j_s i_t} := \begin{cases} c_{ij}, & \text{if } \{i,j\} \in E \\ M, & \text{otherwise} \end{cases}$$

where M is a sufficiently large number.

The node–weights are defined as follows

$$\tilde{b}_{i_s} := \tilde{b}_{i_t} := b_i'/2.$$

Now let $\tilde{x}$ be an optimal solution of the Hitchcock transportation problem on $\tilde{G}$. Then define for $\{i, j\} \in E$

$$x_{ij}^0 := \begin{cases} \tilde{x}_{i_s, j_t} + \tilde{x}_{j_s, i_t}, & \text{if } \tilde{x}_{i_s, j_t} < M \\ 0, & \text{otherwise.} \end{cases}$$

Then x^0 is a maximum cardinality b–matching in G of minimal cost.

Note that with $\tilde{x}$ an optimal HTP–solution also the solution $\hat{x}$ defined by

$$\hat{x}_{i_s, j_t} := \tilde{x}_{j_s, i_t} \quad \text{for} \quad \{i_s, j_t\} \in E$$

is an optimal solution.

ad (ii): Let $(u, v) \in \mathbb{R}^{V_s \cup V_t}$ be an optimal solution to HTP. Then (u, v) fulfills complementary slackness with $\tilde{x}$ as well as with $\hat{x}$.

Now define

$$y_i^0 := (u_{i_s} + v_{i_t})/2 \quad \text{for } i \in V$$

$$y_R^0 := 0 \quad \text{for } R \in \mathcal{S}(b).$$

Then

$$y_i + y_j \leq c_{ij} \quad \text{for} \quad \{i, j\} \in E \quad \text{and}$$

$$y_i + y_j = c_{ij} \quad \text{if} \quad x_{ij}^0 > 0.$$

Hence (x^0, y^0) is a strongly compatible pair with respect to the blossom–characterization and with respect to the cut–characterization of $P^=(G, b)$ as well. Thus $\overline{\mathcal{S}(y^0)} = \emptyset$ and the shrinking family $\mathcal{S}^0 \supset \overline{\mathcal{S}(y^0)}$ can be obtained by applying the three b–matching modification routines of section 14.2. . This requires at most $O(|E|^2)$ steps.

ad (iii): Assume a perfect b–matching x and let x' be the (not necessarily pefect) b'–matching obtained from x by lowering x_e by one unit on at most one edge in $\delta(i)$ for each i with $b_i \neq b_i'$. Then $x'(E) \geq x(E) - |V|$ i.e.

$$d(G, b, x') \le |V|.$$

Now since x^0 is a maximum cardinality b'-matching we also have

$$d(G, b, x^0) \le |V|.$$

$\square$

Thus at most $|V|/2$ applications of the shortest augmenting path procedure are necessary to construct an optimal (perfect) b-matching "starting" from (x^0, y^0, S^0) .

By this "scaling approach" the bMP is solved in at most $O(log \sum_{i \in E} b_i |V| \cdot |E|)$ operations, which is a polynomial in the input–size of this problem.

To achieve this bound the HTP has to be solved by the HTP–scaling algorithm which we have introduced as a clever implementation of the shortest augmenting path concept in section 11.2. . Thus the whole procedure can be viewed as a special implementation of the SAP–approach.

Instead of transforming the b'-matching problem into a HTP one could also apply the scaling–technique on the nonbipartite graph G , directly, thus defining

$$b_i^k = \lfloor b_i/2^k \rfloor \text{ for } 0 \le k \le p - 1, \ i \in V,$$

where $2^{p-1} \le max_{i \in V}\{b_i\} \le 2^p$ and then obtaining an initial b^k-matching x^k "doubling" the optimal extreme b^{k+1}-matching x^{k+1} . Since these degree–constraints are generally not even we may have the problem that $S(b^{k+1})$ and $S(b^k)$ are not related and thus with y^{k+1} the dual solution associated with x^{k+1} we may have

$$y_S^{k+1} > 0 \text{ for } S \notin S(b^k).$$

This results in slightly more complex modification–subroutines to be performed after each "phase".

Marsh [1979] has proposed to perform the scaling–technique on G with altered degree–constraints

$$b_i^k = 2^k \lfloor b_i/2^k \rfloor \text{ for } 0 \leq k \leq p - 1, i \in V ,$$

with $2^{p-1} \leq max_{i \in V}\{b_i\} \leq 2^p$.

Thus b_i^k is obtained by changing to 0 each 1 in the last k "bits" of the binary expansion of b_i .

The advantage of this variant is the fact that for $k \geq 1$ we get $b_i^k \equiv 0 \mod 2$ for all $i \in V$ and hence $S(b^k) = \emptyset$. Thus when having "solved" the b^{k+1}–matching problem the dual solution can be used to initialize the "phase" for solving the b^k–matching problem.

Another variant was recently proposed by Anstee [1983], who first solves the fractional b–matching problem

$$min \ c'x$$

$$Ax = b$$

$$x \geq 0$$

which is essentially a Hitchcock transportation problem on a symmetric bipartite graph. This can be done in $O((log(\sum_{i \in V} b_i) \cdot |E|)$ time. Then by an $O(|E|)$ operation the optimal fractional matching x^f is transformed into an extreme (integer) b–matching x^0 with the property

$$d(G, b, x^0) \leq |V|/3.$$

Moreover the dual solution y^f associated with x^f and $S = \emptyset$ form a strongly compatible triple together with x^0 .

Then Anstee proposes to enter Pulleyblank's blossom–implementation with this structure. Equivalently we could also initialize the SAP–procedure with this strongly compatible triple.

All these variants of the scaling-approach lead to the same bound of $O((\log \sum_{i \in V} b_i) \cdot |V| \cdot |E|)$ operations for the b–matching problem, i.e. are good algorithms in the sense of Edmonds. Thus the b–matching problem and hence the entire class of general matching problems is a well–solved class of integer programming.

References

Aashtiani, H.A. and T.L. Magnanti [1976]: *Implementing primal–dual network flow algorithms.* Research Report OR 055–76, Operations Research Center, Massachusets Institute of Technology, (1976).

Aho, A.V., J.E. Hopcroft and J.D. Ullman [1974]: *The design and analysis of computer algorithms.* Addison–Wesley, Reading (1974).

Ali, A.L., R.V. Helgason, J.L. Kennington and H.S. Lall [1978]: *Primal simplex network codes: State–of–the–art implementation technology.* Networks 8 (1978), pp. 315–339.

Apell, P. [1928]: *Le Problème Géometrique des Deblais et Remblais.* Mem. des Sciences Math. Paris, 27, (1928).

Araoz, J., W.H. Cunningham, J. Edmonds and J. Green–Krotki [1983]: *Reductions to 1–matching polyhedra.* Networks 13 (1983).

Anstee, R.P. [1981]: *An algorithmic proof of Tutte's f–factor theorem.* Research Report CORR 81–28, Dept. of Combinatorics and Optimization, University of Waterloo, Waterloo, Ontario (1981).

Avis, D. and V. Chvatal [1978]: *Notes on Bland's pivoting rule.* Mathematical Programming Study 8 (1978), pp. 24–34.

Bachem, A. [1982]: *Concepts of algorithmic computation.* in: B. Korte (ed.), Modern Applied Mathematics–Optimization and Operations Research, North Holland, Amsterdam (1982), pp.3–49.

Bachem, A. and M. Grötschel [1982]: *New aspects of polyhedral theory.* in: B. Korte (ed.) Modern Applied Mathematics – Optimization and Operations Research, North Holland, Amsterdam (1982), pp. 51–106.

Balinski, M.L. and R.E. Gomory [1964]: *A primal method for the assignment and transportation problems.* Management Science 10 (1964), pp. 578–593.

Balinski, M. [1968]: *Integer programming: Methods, uses, computation.* in: G.B. Dantzig and A.F. Veinott (eds.), Mathematics of the Decision Sciences. Part I. pp. 179–256. American Mathematical Society Providence (1968).

Balinski, M. [1969]: *Labelling to obtain a maximum matching.* in: R.C. Bose and T.A. Bowling (eds.), Combinatorial Mathematics and its Applications, pp. 585–602. The University of North Carolina Press, Chapel Hill (1970).

Balinski, M. [1970]: *On maximum matching, minimum covering and their connection.* in: Proceedings of the Princeton Symposium on Mathematical Programming, pp. 303–311. Princeton University Press (1970).

Balinski, M. [1972]: *Establishing the matching polytope.* Journal of Combinatorial Theory (B) (1972), pp. 1–13.

Balinski, M.L. and A. Russaboff [1974]: *On the assignment polytope*. SIAM Review 16 (1974), pp. 516–525.

Ball, M.O., L. Bodin and R. Dial [1983]: *A matching based heuristic for scheduling mass transit crews and vehicles*. Transportation Science 17 (1983), pp. 4–31.

Ball, M.O. and U. Derigs [1983]: *An analysis of alternate strategies for implementing matching algorithms*. Networks 13 (1983), pp. 517–549.

Barr, R.S., F. Glover and D. Klingman [1977]: *The alternating basis algorithm for assignment problems*. Mathematical Programming 13 (1977), pp. 1–13.

Barr, R.S., F. Glover and D. Klingman [1978]: *The generalized alternating path algorithm for transportation problems*. European Journal of Operations Research 2 (1978), pp. 137–144.

Barr, R.S., F. Glover and D. Klingman [1974]: *An improved version of the out-of-kilter method and a comparative study of computer codes*. Mathematical Programming 7, (1974,) pp. 60–86.

Bartels, R.H. and G.H. Golub [1970]: *The simplex method of linear programming using LU decomposition*. Communications of the ACM 12 (1970), pp. 266–270.

Bazaraa, M.S. and J.J. Jarvis [1977]: *Linear programming and network flows*. John Wiley, New York (1977).

Beale, E.M.L. [1955]: *Cycling in the dual simplex algorithm*. Naval Research Logistics Quaterly 2 (1955), pp. 269–275.

Bellman, R. [1958]: *On a routing problem*. Quart. Apply. Math. 16 (1958), pp. 87–90.

Bennington, G.E. [1973]: *An efficient minimal cost flow algorithm*. Management Science 19 (1973), pp. 1021–1051.

Berge, C. [1957]: *Two theorems in graph theory*. Proc. Natl. Acad. Sci. U. S., 43 (1957) pp. 842–844.

Berge, C [1958]: *Theorie des graphs et ses applications*. Dunod, Paris (1958).

Berge, C. [1958]: *Sur le couplage maximum d'un graphe*. C.R. Acad. Sci Paris 247 (1958), pp. 258–259.

Berge, C. [1962]: *The theory of graphs and its applications*. John Wiley, New York (1962).

Berge, C. [1972]: *Alternating chain methods: A survey*. in: R.C. Read (ed.), Graph Theory and Computing, pp. 1–13, Academic Press, New York – London (1972).

Berge, C. [1973]: *Graphs and Hypergraphs*. North Holland, Amsterdam (1973).

Berge, C. and A. Ghouila–Houri [1965]: *Programming, games and transportation networks*. John Wiley, New York (1965).

Bertsekas, D.P. [1981]: *A new algorithm for the assignment problem*. Mathematical Programming 21 (1981), pp. 152– 171.

Birkhoff, G. [1946]: *Three observations on linear algebra*. Universidad Nacional de Tucuman, Revista Series A 5 (1946), pp. 147–151.

Bland, R.G. [1977]: *New finite pivoting rules for the simplex method*. Mathematics of Operations Research 2 (1977), pp. 103–107.

Bondy, J.A. and U.S.R. Murty [1976]: *Graph theory with applications*. Macmillan, London (1976).

Bradley, G.H. [1975]: *Survey of deterministic networks*. AIIE Transactions 7 (1975), pp.222–234.

Bradley, G.H., G.G. Brown and G.W. Graves [1977]: *Design and implementation of large scale primal transshipment algorithms*. Management Science 24 (1977), pp. 1–34.

Brown, J. [1977]: *Shortest alternating path algorithms*. Networks 4 (1977), pp. 311–334.

Burkard, R.E. and U. Derigs [1980]: *Assignment and matching problems: Solution methods and FORTRAN–Programs*. Lecture Notes in Economics and Mathematical Systems 184, Springer, Berlin (1980).

Busacker, R.G. and P.J. Gowen [1961]: *A procedure for determining a family of minimum–cost network flow patterns*. ORO Technical Report 15, Operations Research Office, Johns Hopkins University (1961).

Busacker, R.G. and T. Saaty [1965]: *Finite graphs and networks*. McGraw–Hill, New York (1965).

Camion, P. [1965]: *Characterization of totally unimodular matrices*. Proc. Amer. Math. Soc., 16 (1965), pp. 1068–1073.

Carpaneto, G. and P. Toth [1980]: *Algorithm 548 (Solution of the assignment problem)*. ACM Transactions on Mathematical Software 6 (1980), pp. 104–111.

Carpaneto, G. and P. Toth [1983]: *Algorithm for the solution of the assignment problem for sparse matrices*. Computing 31 (1983), pp. 83–94.

Charnes, A. and W.W. Cooper [1954]: *The stepping stone method of explaining linear programming calculations in transportation problems*. Management Science 1 (1954), pp. 49–69.

Charnes, A. and W.W. Cooper [1961]: *Management models and industrial applications of linear programming*. Volumes I and II. John Wiley, New York (1961).

Charnes, A., F. Glover, D. Karney, D. Klingman and J. Stutz [1975]: *Past, present, and future of large scale transportation and transshipment computer codes*. Computers and Operations Research 2 (1975), pp. 71–81.

Cherkasky, B.V. [1977]: *Algorithm of construction of maximal flow in networks with complexity $O(|V|^2 \sqrt{|E|})$ operations*. Math. Methods of Solutions of Economic Problems, 7 (1977), pp. 117–125 (in Russian).

Cheung, T.-Y. [1980]: *Computational comparison of eight methods for the maximum network flow problem*. ACM Transactions on Mathematical Software 6 (1980), pp. 1–16.

Christofides, N. [1975]: *Graph theory, an algorithmic approach*. Academic Press, New York (1975).

Chvátal, V. [1973]: *Edmonds polytopes and weakly hamiltonian graphs*. Mathematical Programming 5 (1971), pp. 29–40.

Chvátal, V. [1973]: *Edmonds polytopes and a hierarchy of combinatorial problems*. Discrete Mathematics 4 (1973), pp. 305–337.

Chvátal, V. [1975]: *On certain polytopes associated with graphs*. Journal of Combinatorial Theory B 18 (1975), pp. 138–154.

Conradt, D. [1978]: *Eine effiziente Implementierung des Verfahrens von Edmonds zur Bestimmung eines maximalen Matchings*. TU Berlin, FB 20, Bericht 78–18 (1978).

Conradt, D. und U. Pape [1980]: *Maximales Matching in Graphen*. in: H. Späth (ed.), Ausgewählte Operations Research Software in FORTRAN Oldenbourg, München (1980).

Cruse, A. [1975]: *A note on symmetric doubly–stochastic matrices*. Discrete Mathematics 13 (1975), pp. 109–119.

Cunningham, W.H. [1976]: *A network simplex method*. Mathematical Programming 11 (1976), pp. 105–116.

Cunningham, W.H. [1979]: *Theoretical properties of the network simplex method*. Mathematics of Operations Research 4 (1979), pp. 196–208.

Cunningham, W.H. and J.G. Klincewicz [1983]: *On cycling in the network simplex method*. Mathematical Programming 26 (1983), pp. 182–189.

Cunningham, W.H. and A.B. Marsh, III [1978]: *A primal algorithm for optimum matching*. Mathematical Programming Study 8 (1978), pp. 50–72.

Dantzig, G.B. [1951]: *Application of the simplex method to a transportation problem*. Activity Analysis of Production and Allocation, Cowles Commission Monograph 13, John Wiley, New York (1951), pp. 359–373.

Dantzig, G.B. [1955]: *Upper bounds, secondary constraints, and block triangulation in linear programming*. Econometrica 23 (1955), pp. 174–183.

Dantzig, G.B. [1963]: *Linear programming and extensions*. Princeton University Press, Princeton, New Jersey (1963).

Dantzig, G.B., L.R. Ford and D.R. Fulkerson [1956]: *A primal–dual algorithm for linear programs*. pp. 171–181 in: H. W. Kuhn and A. W. Tucker (eds.), Linear Inequalities and Related Systems, Princeton University Press, Princeton, N.J. (1956).

Dantzig, G.B., A. Orden and P. Wolfe [1955]: *A generalized simplex method for minimizing a linear form under linear inequalities*. Pacific Journal of Mathematics 5(1955), pp. 183–195.

Darby–Dowman, K. [1980]: *The exploition of sparsity in large scale linear programming problems–Data structures and restructering algorithms for basis matrices*. Ph. D. Thesis, Brunel University, Uxbridge, England (1980).

Darby–Dowman, K. and G. Mitra [1981]: *An investigation of algorithms used in the restructering of linear programming basis matrices prior to inversion*. in: Hansen, P. (ed.), Studies on graphs and discrete programming, Annals of Discrete Mathematics 11, North Holland, Amsterdam (1981), pp. 69–93.

Deming, R.W. [1979]: *Independence numbers of graphs – an extension of the König–Egerváry theorem*. Discrete Mathematics 27 (1979), pp. 23–33.

Derigs, U. [1979]: *A generalized hungarian method for solving minimum weight perfect matching problems with algebraic objective*. Discrete Applied Mathematics 1 (1979), pp. 167–180.

Derigs, U. [1980]: *The cardinality matching problem – Methods and computations*. in: P. Hansen and D. de Werra (eds.), Regards sur la theory des graphes, Lausanne (1980), pp. 199–203.

Derigs, U. [1981]: *A shortest augmenting path method for solving minimal perfect matching problems*. Networks 11 (1981), pp. 379–390.

Derigs, U. [1982]: *Shortest augmenting paths and sensitivity analysis for optimal matchings*. Report 82222 OR, Institut für Ökonometrie und Operations Research, Universität Bonn (1982).

Derigs, U. [1983]: *Optimale Zuordnungen und Matchings: Anwendungen und Verfahren*. in: W. Bühler et al. (eds.), Operations Research Proceedings 1983, Berlin (1983), pp. 335–344.

Derigs, U. and H. Heske [1980]: *A computational study on some methods for solving the cardinality matching problem*. Angew. Inf. Appl. Inf. 22 (1980), pp. 149–154.

Derigs, U. and G. Kazakidis [1980]: *On two methods for solving minimal perfect matching problems*. in: J. Karup and S. Walukiewicz (eds.), Proceeding of the Second Danish/Polish Mathematical Programming Seminar (1980), pp. 85–100.

Devine, M. and F. Glover [1972]: *Computational study of the symmetric assignment problem*. Working Paper, School of Industrial Engineering, University of Oklahoma, Norman (1972).

Dial, R. [1969]: *Algorithm 360: Shortest path forest with topological ordering*. [H], Communications of the ACM 12 (1969), pp. 632–633.

Dial, R., F. Glover, D. Karney and D. Klingman [1979]: *A computational analysis of alternative algorithms and labeling techniques for finding shortest path trees*. Networks 9 (1979), pp. 215–248.

Dijkstra, E.W. [1959]: *A note on two problems in connection with graphs*. Numerische Mathematik 1 (1959), pp. 269–271.

Dinic, E.A. [1970]: *Algorithm for solution of a problem of maximum flow in a network with power estimation*. Sov. Math. Dokl. 11 (1970), pp. 1277–1280.

Dinic, E. and M. Kronrod [1969]: *An algorithm for the solution of the assignment problem*. Sov. Math. Dokl.10 (1969).

Dörfler, W. and J. Mühlbacher [1972]: *Bestimmung eines maximalen Matching in beliebigen Graphen*. Computing 9 (1972), pp. 251– 257.

Dörfler, W. and J. Mühlbacher [1974]: *Ein verbesserter Matching Algorithmus*. Computing 13 (1974), pp. 389–397.

Domschke, W. [1973]: *Two new algorithms for minimal cost flow problems*. Computing 11 (1973), pp. 275–285.

Dorhout, B. [1973]: *Het lineare toewijzungsproblem, vergelijken van algoritmen*. Report BN 21 Stichting Mathematisch Centrum, Amsterdam (1973).

Dorhout, B. [1977]: *Experiments with some algorithms for the linear assignment problem*. Mathematisch Centrum, Working Paper, BW 39 (1977).

Edmonds, J. [1965a]: *Path, trees, and flowers*. Canadian Journal of Mathematics 17 (1965), pp. 449–467.

Edmonds, J. [1965b]: *Maximum matching and a polyhedron with 0, 1 vertices*. J. Res. NBS, 69B (1965), pp. 125–130.

Edmonds, J. [1967]: *An introduction to matching*. mimeographed notes, Engineering Summer Conference, The University of Michigan, Ann Arbor (1967).

Edmonds, J. [1971]: *Matroids and the greedy algorithm*. Mathematical Programming 1 (1971), pp. 127–136.

Edmonds, J. and E.L. Johnson [1970]: *Matching: A well–solved class of integer linear programs*. in: R. Guy (ed), Combinatorial Structure and Their Applications, Gordon and Breach, New York (1970), pp. 89–92.

Edmonds, J. and E.L. Johnson [1973]: *Matching, Euler tours and the chinese postman*. Mathematical Programming 5 (1973), pp. 88–124.

Edmonds, J. and R.M. Karp [1972]: *Theoretical improvements in algorithmic efficiency for network flow problems.* Journal of the ACM 19 (1972), pp. 248–264.

Edmonds, J. and E. Koch [1981]: *Simplex methods for optimum matching problems.* Technical Report 109, IBM Israel Scientific Center (1981).

Edmonds, J., E.L. Johnson and S. Lockhart [1969]: *Blossom I, a code for matching.* unpublished Report, IBM T.J. Watson Research Center, Yorktown Heights, New York (1969).

Edmonds, J. and W. Pulleyblank [1974]: *Facets of 1–matching polyhedra.* in: Hypergraph Seminar. Lecture Notes in Mathematics 411 (1974), pp. 214–242.

Egerváry, J. [1931]: *Matrixok kombinatorius tulajdonságairól.* Mat. Fiz. Lapok 38 (1931), pp. 16–28.

Elias, P., A. Feinstein and C.E. Shannon [1956]: *Note on maximum flow through a network.* IRE Trans. on Information Theory, IT–2 (1956), pp. 117–119.

Elmaghraby, S. [1970]: *Some network models in operations research.* Springer-Verlag, New York (1970).

Engquist, M. [1980]: *A successive shortest path algorithm for the assignment problem.* INFOR 20 (1982), pp. 370–384.

Even, S. and O.Kariv [1975]: *An $0(n^{2.5})$ algorithm for maximum matching in general graphs.* Proc. 16th Annual Symp. on Foundation of Computer Science, IEEE, New York (1975), pp. 100–112.

Flood, M.M. [1960]: *An alternative proof of a theorem of König as an algorithm for the Hitchcock problem.* in: R. Bellman and M. Hall, Jr. (eds.), Proceedings of Symposia in Applied Mathematics X. Combinatorial analysis. Amer. Math. Soc. Providence (1960).

Florian, M. and M. Klein [1970]: *An experimental evaluation of some methods of solving assignment problem.* CORS Journal 8 (1970).

Ford, L.R. [1956]: *Network flow theory.* The Rand Corp., P–923, August (1956).

Ford, L.R. and D.R. Fulkerson [1956]: *Maximal flow through a network.* Canadian Journal of Mathematics 8 (1956), pp. 399–404.

Ford, L.R. and D.R. Fulkerson [1957a]: *A primal–dual algorithm for the capacitated Hitchcock problem.* Naval Research Logistics Quarterly 4 (1957), pp. 47–54.

Ford, L.R. and D.R. Fulkerson [1957b]: *A simple algorithm for finding maximal network flows and an application to the Hitchcock Problem.* Canadian Journal of Mathematics 9 (1957), pp. 210–218.

Ford, L.R. and D.R. Fulkerson [1962]: *Flows in networks.* Princeton University Press, Princeton, New Jersey (1962).

Fulkerson, D.R. [1961]: *An out-of-kilter method for minimal-cost flow problems*. SIAM Journal on Applied Mathematics 9 (1961), pp. 18–27.

Fulkerson, D.R. [1966]: *Flow networks and combinatorial operations research*. American Mathematical Monthly 73 (1966), pp. 115–138.

Gabow, H.N. [1975]: *An efficient implementation of Edmonds' maximum matching algorithm*. Journal of the ACM 23 (1975), pp. 221–234.

Gabow, H.N. [1983]: *An efficient reduction technique for degree– constrained subgraph and bidirected network flow problems*. Proceedings of the 15th Annual Symposium on Theory of Computing, (1983), pp. 448–456.

Gabow, H.N. and R.E. Tarjan [1983]: *A linear time algorithm for a special case of disjoint union*. Proceedings of the 15th Annual ACM Symposium on Theory of Computing (1983), pp. 246–251.

Gale, D. [1957]: *A theorem on flows in networks*. Pacific Journal of Mathematics 7 (1957), pp. 1073–1082.

Galil, Z. [1978]: *A new algorithm for the maximal flow problem*. Proc. 19th Symposium on Foundations of Computer Science, (1978), pp. 231–245.

Galil, Z. [1981]: *On the theoretical efficiency of various network flow algorithms*. Theoretical Computer Science 14 (1981), pp. 103–111.

Galil, Z., S.Micali and H. Gabow [1982]: *Priority queues with variable priority and an $O(|E||V|\log|V|)$ algorithm for finding a maximal weighted matching in general graph*. Proc. 23rd Annual Symposium on Foundation of Computer Science (1982), pp. 255–261.

Galil, Z. and A. Naamad [1979]: *Network flow and generalized path compression*. Proc. 11th Annual ACM Symposium on Theory of Computing, (1979), pp. 13–26.

Gallo, G. and S. Pallotino [1983]: *Shortest path methods: A unifying approach*. presented at: Netflow 83, an International Workshop on Network Flow Optimization Theory and Practise, Pisa, Italy (1983), pp. 28–31.

Gallo, G., S. Pallotino and C.R. Ruggeri [1982]: *On the experimental efficiency of shortest paths*. C.N.R. – P.F. Informatica – SOFMAT 27 (1982).

Garey,M.R. and D.S. Johnson [1979]: *Computers and intractibility: A guide to the theory of NP-completeness*. Freeman, San Francisco (1979).

Gass, S.I. [1975]: *Linear Programming – Methods and Applications*. McGraw Hill , New York (1975).

Gassner, B.J. [1964]: *Cycling in the transportation problem*. Naval Research Logistic Quarterly 11 (1964), pp. 43–58.

Gavish, B. and P. Schweitzer [1974]: *An algorithm for combining truck trips*. Transportation Science 8 (1974), pp. 13–23.

Gavish, B., P. Schweitzer and E. Shlifer [1977]: *The zero pivot phenomenon in transportation and assignment problems and its computational implications.* Mathematical Programming 12 (1977), pp. 226–240.

Gilsinn, J. and C. Witzgall [1973]: *A performance comparison of labeling algorithms for calculating shortest path trees.* Technical Note 772, National Bureau of Standards, Washington, D.C. (1973).

Gleyzal, A.N [1955]: *An algorithm for solving the transportation problem.* J. Res. NBS 54 (1955), pp. 123–216.

Glicksman, S., L. Johnson and L. Eselson [1960]: *Coding the transportation problem.* Naval Research Logistics Quarterly 7 (1960), pp. 169–183.

Glover, F. [1967]: *Minimum complete matchings.* ORC– Report 67–15, University of California, Berkely (1967).

Glover, F., Glover, R. and D. Klingman [1983]: *Threshold assignment algorithm.* Working Paper (1983).

Glover, F., D. Karney and D. Klingman [1974]: *Implementation and computational comparisons of primal, dual and primal–dual computer codes for minimum cost flow network flow problems.* Networks 4 (1974), pp. 191–212.

Glover, F., D. Karney, D. Klingman and A. Napier [1974]: *A computational study on start procedures, basis change criteria and solution algorithms for transportation problems.* Management Science 20 (1974), pp. 793–813.

Glover, F., D. Klingman, J. Note and D. Whitman [1979]: *Comprehensive computer evaluation and enhancement of maximum flow algorithms.* Research Report 356, Center of Cybernetic Studies, The University of Texas, Austin (1979).

Glover, F., D. Klingman and N. Phillips [1982]: *Partitioning shortest path algorithm.* Research Report, University of Texas, Austin, TX (1982).

Goldman, A.J. [1964]: *Optimal matchings and degree constrained subgraphs.* J. Res. NBS 68B (1964), pp. 27–29.

Gondran, M. [1974]: *Remarque sur le polytope des couplages d'Edmonds.* Discrete Mathematics 9 (1974), pp. 235–238.

Gondran, M. and M. Minoux [1979]: *Graphes et algorithmes.* Editions Eyrolles (1979).

Green–Krótki, J.J. [1980]: *Matching polyhedra.* M. Sc. Thesis, Carleton University (1980).

Grigoriadis, M. and T. Hsu [1979]: *The rutgers minimum cost network flow subroutine.* Sigmap 26 (1979), pp. 17–18.

Grötschel, M. [1977]: *Polyhedrische Charakterisierungen Kombina- torischer Optimierungsprobleme.* Dissertation, Universität Bonn (1977), Verlag A. Hain, Meisemheim (1977).

Grötschel, M., L.Lovász and A. Schrijver [1981]: *The ellipsoid method and its consequences in combinatorial optimization.* Combinatorica 1 (1981), pp. 169–197.

Hall, M. [1956]: *An algorithm for distinct representatives.* Amererican Mathematical Monthly 63 (1956), pp. 716–717.

Hall, P. [1935]: *On representatives of subsets.* J. London Math. Soc. 10 (1935), pp. 26–30.

Harary, F. [1969]: *Graph theory.* Addison–Wesley, Reading (1969).

Hasselström, D. [1976]: *Connecting bus–routes at a point of intersection.* Paper presented at the Second European Congress on Operations Research, Stockholm, Schweden, (1976).

Hasselström, D. [1981]: *Public transportation planning– A mathematical programming approach.* Ph. D. Thesis, Göteborg, Schweden (1981).

Hassin, R. [1983]: *The minimum cost flow problem: A unifying approach to dual algorithms and a new tree–search algorithm.* Mathematical Programming 25 (1983), pp. 228–239.

Havel, T.F. [1975]: *The combinatorial distance geometry approach to the calculation of modular confirmation.* Ph. D. Thesis, University of California (1975).

Heller, I. and G.B. Tompkins [1956]: *An extension of a theorem of Dantzig's.* in: H.W. Kuhn et al. (eds.), Linear Inequalities and Related Systems, Princeton University Press, N.J. (1956), pp. 247–252.

Heske, A. [1978]: *Die Bestimmung maximaler Matchings in beliebigen Graphen.* Diploma Thesis, Mathematisches Institut, Universität zu Köln (1978).

Hitchcock, F.L. [1941]: *The distribution of products from several sources to numerous localities.* Journal of Mathematics and Physics 20 (1941), pp. 224–230.

Hoffman, A.J. [1960]: *Some recent applications of the theory of linear inequalities to extremal combinatorial analysis.* Proc. Symposium on Appl. Math. 10 (1960), pp. 113–127.

Hoffman, A.J. and J.B. Kruskal [1956]: *Integral boundary points of convex polyhedra.* in H.W. Kuhn and A.W. Tucker (eds.), Linear inequalities and related systems, Annals of Mathematics Study No. 38, Princeton Univ. Press, Princeton, New Jersey (1956), pp. 233–246.

Hoffman, A.J. and H.W. Kuhn [1956]: *Systems of distinct representatives and linear programming.* American Mathematical Monthly 63 (1956), pp. 455–460.

Hoffman, A.J. and H.M. Markowitz [1963]: *A note on shortest path, assignment, and transportation problems.* Naval Research Logistics Quart. 10 (1963), pp. 375–380.

Holland, O. [1983]: *Optimierung kapizitierter und nicht–kapizitierter b–Matchings unter Verwendung der Linearen Programmierung.* Diploma Thesis, University of Bonn (1983).

Hopcroft, J.E. and R.M. Karp [1973]: *An $n^{5/2}$ algorithm for maximum matchings in bipartite graphs.* SIAM Journal of Computing 2 (1973), pp. 225–231.

Hu, T.C. [1969]: *Integer programming and network flows.* Addison–Wesley, Reading, (1969).

Hung, M.S. [1983]: *A polynomial simplex method for the assignment problem.* Operations Research 31 (1983), pp. 595–600.

Hung, M.S. and W.O. Rom [1980]: *Solving the assignment problem by relaxation.* Operations Research 28 (1980), pp. 969–982.

Iri, M. [1960]: *A new method of solving transportation–network problems.* J. Op. Res. Soc. Japan, 3 (1960), pp. 27–87.

Itai, A. and Y. Shiloach [1979]: *Maximum flow in planar networks.* SIAM Journal of Computing 8 (1979), pp. 135–150.

Jenkins, T.A. [1974]: *Matchoids: a generalization of matchings and matroids.* Ph. D. Thesis, University of Waterloo, Waterloo, Ontario (1974).

Jensen, P.A. and J.W. Barnes [1980]: *Network flow programming.* Wiley, New York (1980).

Jewell, W.S. [1958]: *Optimal flow through networks.* Interim Technical Report no. 8, Massachusetts Institute of Technology (1958).

Jewell, W.S. [1962]: *Optimal flow through networks with gains.* Operations Research 10 (1962), pp. 476–499.

Johnson, E.L. [1965]: *Networks, graphs and integer programming.* Ph. D. Thesis, issued as Report ORC 65–1, Operations Research Center, The University of California, Berkeley (1965).

Johnson, E.L. [1966]: *Networks and basic solutions.* Operations Research 14 (1966), pp. 619–623.

Johnson, E.L. [1972]: *On shortest paths and sorting.* Proc. of the 1972 ACM Conference, Boston, (1972), pp. 510–517.

Johnson, E.L. [1978]: *Flows in Networks in: S.E. Elmaghraby and J.J. Moder (eds.), Handbook of Operations Research, Foundations and Fundamentals.* Van Nostrand Reinhold, New York (1978).

Kameda, T. and I. Munroe [1974]: *A $O(|V| \cdot |E|)$ algorithm for maximum matching of graphs.* Computing 12 (1974), pp. 91–98.

Kantorovich, L. [1942]: *On the translocation of masses.* Compt. Rend. (Doklady) Acad. Sci., 37 (1942), pp. 199–201.

Kantorovich, L. and M.K. Gavurin [1949]: *The application of mathematical methods to problems of freight flow analysis*. Akademia Nauk SSSR, (1949).

Kariv, O. [1977]: *An $0(n^{2.5})$ algorithm for maximum matching in general graphs*. Ph. D. Thesis, Technion Haifa, Israel (1977).

Karp, R.M. [1972]: *Reducibility among combinatorial problems*. in: R.E. Miller and J.W. Thatcher (eds.), Complexity of Computer Computations, Pergamon Press, Oxford and New York (1972), pp. 85–103.

Karp, R.M. [1975]: *On the computational complexity of combinatorial problems*. Networks 5 (1975), pp. 45–68.

Karzanov, A.V. [1974]: *Determining the maximal flow in a network with the method of preflows*. Soviet Math. Dokl., 15 (1974), pp. 434–437.

Kastning, C. (editor) [1976]: *Integer Programming and related areas:*. A Classified Bibliography, Lecture Notes in Economics and Mathematical Systems 128, Springer, Berlin (1976).

Katz, M. [1970]: *On the extreme points of a certain convex polytope*. Jornal of Combinatorial Theory 8 (1970), pp. 417–423.

Kennington, J.L. and R.V. Helgason [1980]: *Algorithms for network programming*. John Wiley, New York, 1980.

Khachian, L.G. [1979]: *A polynomial algorithm in linear programming*. Doklady Academia Nauk SSSR 244 (1979), pp. 1093–1096.

Klee, V. and C. Witzgall [1968]: *Facets and vertices of transportation polytopes*. in: George B. Dantzig and Arthur F. Veinott, Jr. (eds.), Mathematics of the Decision Sciences II, American Mathematical Society, Providence, R.I (1968), pp. 257–282.

Klein, M. [1967]: *A primal method for minimal cost flows with application to the assignment and transportation problems*. Management Science 14 (1967), pp. 205–220.

Klingman, D., A. Napier and J. Stutz [1974]: *NETGEN–A program for generating large–scale (un)capacitated assignment, transportation, and minimum cost flow network problems*. Management Science 20 (1974), pp. 814–821.

Koch, E. [1981]: *Ramifications of matching theory*. Ph. D. Thesis, Waterloo (1981).

König, D. [1931]: *Graphok és matrixok*. Mat. Fiz. Lapok 38 (1931), pp. 116–119.

König, D. [1936]: *Theorie der endlichen und unendlichen Graphen*. Leipzig, Akademische Verlagsgesellschaft (1936).

Koopmans, T.C. [1949]: *Optimum utilization of the transportation system.* Proceedings of the International Statistical Conferences, Washington, D.C. (1947), published in: Scientific Papers of Tjalling C. Koopmans, Springer–Verlag, New York (1970), pp. 184–193.

Koopmans, T.C. [1951]: *Activity analysis of production and allocation.* Cowles Commission Monograph N. 13, John Wiley, New York (1951).

Koopmans, T.C. and S. Reiter [1951]: *A model of transportation.* Activity analysis of production and allocation, Cowles Commission Monograph 13 (1951), Wiley, pp. 222–259.

Korte, B. and L. Lovász [1981]: *Mathematical structures underlying greedy algorithms.* in: F. Gécseg (ed.), Fundamentals of Computation Theory. Lecture Notes in Computer Science 117, Springer Berlin, Heidelberg, New York (1981), pp. 205–209.

Korte, B. and L. Lovász [1984]: *Greedoids, a structural framework for the greedy algorithm.* to appear in: W.R. Pulleyblank (ed.), Progress in Combinatorial Optimization.Proceedings of the Silver Jubilee Conference on Combinatorics, Waterloo, 1982, Academic Press, London, New York, San Francisko (1984)

Kuhn, H.W. [1955]: *The Hungarian method for the assignment problem.* Naval Research Logistics Quarterly 2 (1955), pp. 83–97.

Kuhn, H.W. [1956]: *Variants of the Hungarian method for assignment problems.* Naval Research Logistics Quarterly 3 (1956), pp. 253–258.

Lawler, E.L. [1971]: *Matroids with parity conditions, a new class of combinatorial optimization problems.* Memorandum ERL– M334, University of California, Berkeley (1971).

Lawler, E.L. [1976]: *Combinatorial optimization: Networks and matroids.* Holt, Rinehart and Winston (1976).

Lovász, L. [1975]: *Three short proofs in graph theory.* Journal of Combinatorial Theory (B) 19 (1975), pp. 269–271.

Lovász, L. [1979]: *Graph theory and integer programming.* Annals of Discrete Mathematics 4 (1979), pp. 141–158.

Lovász, L. [1980a]: *Selecting independent lines from a family in a space.* Acta Sci. Math. (Szeged) 42 (1980), pp. 121–131.

Lovász, L. [1980b]: *Matroid matching and some applications.* Journal of Combinatorial Theory (B) 28 (1980), pp. 208–236.

Lovász, L. [1981]: *The matroid matching problem.* in: L. Lovász and V. Sós (eds.), Algebraic methods in graph theory, North–Holland, Amsterdam (1981), pp. 495–517.

Malhotra, V.M., M.P. Kumar and S.N. Maheshwari [1978]: *An $O(|V|^3)$ algorithm for finding maximum flows in networks*. Inf. Proc. Letters 7 (1978), pp. 277–278.

McGinnis, L.F. [1983]: *Implementation and testing of a primal–dual algorithm for the assignment problem*. Operations Research 31 (1983), pp.277–291.

Marsh, A.B. [1979]: *Matching algorithms*. Ph. D. Thesis, Johns Hopkins Univ., Baltimore (1979).

Micali, S. and V.V. Vazirani [1980]: *An $O(\sqrt{|V|} \cdot |E|)$ algorithm for finding maximum matching in general graphs*. Proc. 21st Annual Symposium on Foundation of Computer Science, Long Beach, California: IEEE (1980), pp. 17–27.

Minty, G.J. [1957]: *A comment on a shortest route problem*. Operations Research 5 (1957), p. 724.

Minty, G.J. [1958]: *A variant of the shortest–route problem*. Operations Research 6 (1958), p. 882.

Minty, G.J. [1960]: *Monotone networks*. Proc. Roy. Soc. London, Ser. A 257 (1960), pp. 194–212.

Monge [1781]: *Déblai et Remblai*. Mem. Avad. Sci., 1781.

Moore, E.F. [1959]: *The shortest path through a maze*. Proc. Int. Symp. on the Theory of Switching, Part II, April 2–5, 1957. Also published in: The Annals of the Computation Laboratory of Harvard University 30, Harvard University Press (1959), pp. 285–292.

Motzkin, T.S. [1956]: *The assignment problem*. Proc. of Symposia in Applied Mathematics Vol. VI–Numerical Analysis, McGraw–Hill, New York (1956).

Mulvey, J.M. [1978a]: *Pivot strategies for primal–simplex network codes*. Journal of the ACM 25 (1978), pp. 266–270.

Mulvey J.M. [1978b]: *Network relaxation for set covering, set partitioning and other integer programming problems*. Research Report, University of Texas, Center of Cybernetic Studies(1978).

Mulvey, J.M. [1978]: *Testing of a large–scale network optimization program*. Mathematical Programming 15 (1978), pp. 291–314.

Munkres, J. [1957]: *Algorithm for assignment and transportation problem*. Journal of SIAM 5 (1957), pp. 32–38.

Murty, K.G. [1967]: *The symmetric assignment problem*. Report ORC 67–12, University of California, Berkely (1967).

Nemhauser, G. and G. Weber [1979]: *Optimal set partitioning matchings and Lagrangian Duality*. Naval Research Logistics Quarterly 26 (1979), pp. 553–563.

Noltemeier, H. [1976]: *Graphentheorie mit Algorithmen und Anwendungen*. De Gruyter, Berlin – New York (1976).

Norman, R.Z. and M.O. Rabin [1959]: *An algorithm for a minimum cover of a graph*. Proc. Amer. Math. Soc. 10 (1959), pp. 315–319.

Orden, A. [1956]: *The trans–shipment problem*. Management Science 3 (1956), pp. 276–285.

Ore,O. [1962]: *Theory of graphs*. American Mathematical Society, Providence, Rhode Island (1962).

Padberg, M. [1975]: *Characterizations of totally unimodular, balanced and perfect matrices*. in: B. Roy (ed.) Combinatorial Programming: Methods and Applications, Reidel, Boston (1975), pp. 275–284.

Padberg, M.W. and M.R. Rao [1982]: *Odd minimum cut–sets and b–matchings*. Mathematics of Operations Research 7 (1982), pp. 67– 80.

Papadimitriou, C.H. and K. Steiglitz [1982]: *Combinatorial Optimization: Algorithms and Complexity*. Prentice Hall, Englewood Cliffs. NJ, (1982).

Pape, U. [1974]: *Implementation and efficiency of Moore–algorithms for the shortest route problem*. Mathematical Programming 7 (1974), pp. 212–222.

Potts, R.and R. Oliver [1972]: *Flows in transportations networks*. Academic Press, New York (1972).

Pulleyblank, W. [1973]: *Faces of matching polyhedra*. Ph.D.Thesis, Faculty of Mathematics, University of Waterloo (1973).

Pulleyblank, W.R. [1980]: *Dual integrality in b– matching problems*. Mathematical Programming Study 12 (1980), pp. 176–196.

Pulleyblank, W.R. [1981]: *Total dual integrality and b–matchings*. Operations Research Letters 1 (1981), pp. 28–30.

Pulleyblank, W.R. [1983]: *Polyhedral Combinatorics*. in: A. Bachem et al. (eds.), Mathematical Programming–The state of the art (Bonn 1982), Berlin (1983), pp. 312–345.

Roberts, A.B. [1969]: *Minimum complete matchings*. Thesis, University of Texas at Austin (1969).

Roohy–Laleh, E. [1981]: *Improvements of the theoretical efficiency of the network simplex method*. M. Sc. Thesis, Ottawa (1981).

Schrijver, A. [1981]: *Short proofs on the matching polyhedron.* Journal of Combinatorial Theory (B) 34 (1983), pp. 104–108.

Schrijver, A. and P.D. Seymour [1977]: *A proof of total dual integrality of matching polyhedra*. Report ZN 79/77, Math. Centrum, Amsterdam (1977).

Seymour, P.D. [1979]: *Sums of circuits*. in: Bondy, J.A. et al (eds.), Graph Theory and related Topics, Academic Press, New York (1979), pp. 341–355.

Shapiro, J.F. [1977]: *A note on the primal–dual and out–of–kilter algorithms for network optimization problems.* Networks 7 (1977), pp. 81–88.

Shiloach, Y. [1978]: An $O(n \cdot I \log^2 I)$ maximum–flow algorithm. Tech. Report STAN–CS–78–802, Computer Science Dept., Stanford University (1978).

Silver, R. [1960]: *An algorithm for the assignment problem.* Communications of the ACM 3 (1960), pp. 605–606.

Silver, R. [1960]: *Algorithm 27.* Assignment, Communications of the ACM 3 (1960), pp. 603–604.

Simmonard, Michel [1966]: *Linear Programming (William S. Jewell, trans.).* Prentice–Hall (1966).

Sleator, D.D. [1980]: *An $O(nm \log n)$ algorithm for maximum network flow.* Technical Report STAN–CS–80–831 Computer Science Department, Stanford University (1980).

Smith, L.W. [1956)]: *Current status of the industrial use of linear programming.* Management Science 2 (1956), pp. 156–158.

Srinivasan, V. and G.L. Thompson [1972]: *Accelerated algorithms for labeling and relabeling of trees with application for distribution problems.* Journal of the ACM 19 (1972), pp. 712–726.

Srinivasan, V. and G.L. Thompson [1973]: *Benefit–cost analysis of coding techniques for the primal transportation algorithm.* Journal of the ACM (1973), pp. 194–213.

Sterboul, F. [1979]: *A characterization of the graphs in which the transversal number equals the matching number.* Journal of Combinatorial Theory (B) 27 (1979), pp. 228–229.

Tabourir, Y. [1972]: *Un algorithme pour le problème d'affectation.* RAIRO 6 (1972), pp. 3–15.

Thakker, B.G. [1972]: *Matchings and weighted graphs.* Report, Imperial College of Science and Technology, London (1972).

Thompson, G.L. [1981]: *A recursive method for solving assignment problems.* in: Hansen, P. (ed), Studies on Graphs and Discrete Programming, Annals of Discrete Mathematics 11 North–Holland, Amsterdam, (1981), pp. 319–343.

Thompson, G.L. [1983]: *Comparison of two dimension expanding assignment algorithms.* Methods of Operations Research 45 (1983), pp. 357–365.

Tomizawa, N. [1971]: *On some techniques useful for solution of transportation network problems.* Networks 1 (1971), pp. 173–194.

Tutte, W.T. [1947]: *The factorization of linear graphs.* J. London Math. Soc. 22 (1947), pp. 107–111.

Tutte, W.T. [1952]: *The factors of graphs.* Canadian Journal of Mathematics 4 (1952), pp. 314–328.

Tutte, W.T. [1954]: *A short proof of the factor theorem for finite graphs.* Canadian Journal of Mathematics 6 (1954), pp. 347–352.

Tutte, W.T. [1981]: *Graph factors.* Combinatorica 1 (1981), pp. 79–97.

Urquhart, R.J. [1967]: *Degree–constrained subgraphs of linear graphs.* Ph. D. Thesis, The University of Michigan, Ann Arbor (1967).

Wagner, H.M. [1959]: *On a class of capacitated transportation problems.* Management Science 5 (1959), pp. 304–318.

Weber, G.M. [1978]: *A solution technique for binary integer programming using matchings on graphs.* Ph. d. Thesis, Cornell University (1981).

Weber, G. [1981]: *Sensitivity analysis of optimal matchings.* Networks 11 (1981), pp. 41–56.

Weintraub, A. and F. Barahona [1979]: *A dual algorithm for the assignment problem.* Publication 79/02/C, Departmento de Industrias, Universidad de Chile–Sede Occidente, Santiago, Chile (1979).

Weyl, H. [1935]: *Elementare Theorie der konvexen Polyheder.* Commentarii Mathematici Helvetici 7 (1935), pp. 290–306.

White, L.J. [1969]: *A parametric study of matchings and coverings in weighted graphs.* Technical Report, The University of Michigan, (1969).

Witzgall, C. and C.T. Zahn, Jr. [1965]: *Modification of Edmonds' algorithm for maximum matching of graphs.* J. Res. NBS 69B (1965), pp. 91–981.

Yakovleva, M.A. [1959]: *Problem of minimum transportation expense.* in: V.S. Nemchinov (ed.), Applications of Mathematics to Economics Research, Moscow (1959), pp. 390–399.

Zadeh, N. [1972]: *Theoretical efficiency of the Edmonds–Karp algorithm for computing maximal flows.* Journal of the ACM 19 (1972), pp. 217–224.

Zadeh, N. [1973]: *A bad network problem for the simplex method and other minimum cost flow algorithms.* Mathematical Programming 5 (1973), pp. 255–266.

Zadeh, N. [1973]: *More pathological examples for network flow problems.* Mathematical Programming, 5 (1973), pp. 217–224.

Zadeh, N. [1979]: *A simple alternative to the out–of–kilter algorithm.* Technical Report 25, Stanford University, Department of Operations Research (1979).

Zadeh, N. [1980]: *Near–equivalence of network flow algorithms.* Working Paper, Stanford University (1980).